Industrial and Applied Mathematics

Editor-in-chief

Abul Hasan Siddiqi, Greater Noida, India

The Industrial and Applied Mathematics series publishes high-quality research-level monographs, lecture notes and contributed volumes focusing on areas where mathematics is used in a fundamental way, such as industrial mathematics, bio-mathematics, financial mathematics, applied statistics, operations research and computer science.

More information about this series at http://www.springer.com/series/13577

Priti Kumar Roy

Mathematical Models for Therapeutic Approaches to Control HIV Disease Transmission

 Springer

Priti Kumar Roy
Department of Mathematics
Jadavpur University
Kolkata
India

ISSN 2364-6837   ISSN 2364-6845   (electronic)
Industrial and Applied Mathematics
ISBN 978-981-287-851-9   ISBN 978-981-287-852-6   (eBook)
DOI 10.1007/978-981-287-852-6

Library of Congress Control Number: 2015950916

Springer Singapore Heidelberg New York Dordrecht London

Printed on acid-free paper

Springer Science+Business Media Singapore Pte Ltd. is part of Springer Science+Business Media
(www.springer.com)

Three decades into this crisis, let us set our sights on achieving the three zeros—zero new HIV infections, zero discrimination and zero AIDS-related deaths.

—Secretary-General of the United Nations Ban Ki-moon, on the World AIDS Day, 2010

Prevention, treatment and care are now saving millions of lives not only in the world's richest countries but also in some of the world's poorest countries. And for many, with testing and access to the right treatment, the disease that was once a death sentence now comes with a good chance at a healthy and productive life. And that's an extraordinary achievement....

We will stand with you every step of this journey until we reach the day that we know is possible, when all men and women can protect themselves from infection; a day when all people with HIV have access to the treatments that extend their lives; the day when there are no babies being born with HIV or AIDS, and when we achieve, at long last, what was once hard to imagine—and that's an AIDS-free generation.

—President of the United States of America Barrack Obama, on the World AIDS Day, 2013

This book is dedicated to my family and the pandemic victims suffering from AIDS. Acquired Immunodeficiency Syndrome shares the limelight for quite a few decades due to its evolving nature. Now, new treatment policies has changed the casualties of AIDS victims in a dramatic way. Society can now coexist with AIDS, but we can ask the next question 'Can the disease be actually controlled or eradicated?' Can we minimize the disease transmission or progression through combination of different drug therapies? Can impulsive drug therapy lead us to a safe clinical regime and washout the disease with optimal control approach? Can it serve as a new era in clinical methodology? In this book we are searching the resolution through mathematical outlook. We hope that our findings can enlight the world from dawn to dusk in combating the demonic disease.

Preface

Human immunodeficiency virus (HIV), causing impairment of human immune system and inflicting the disease Acquired Immune Deficiency Syndrome (AIDS), is a grave problem that the human race encounters and needs immediate attention to formulate potential treatment strategy against the disease. The HIV epidemic in India has a major impact on the overall spread of HIV in Asia, the Pacific, and around the world. India is the second country only after South Africa in terms of the overall number of people living with the disease. HIV virus has frequent mutation and an infected individual often harbors many variations. The high mutation rate of HIV allows developing resistance against the drugs used to treat it. Development of treatments and vaccines depends not only on knowledge of the complex life cycle of virus, but also on understanding of the difficulty in the management of immune system.

To control HIV, development of the medicines and vaccines will be required in near future. However, poverty and politics are the major obstacles to fight against the disease. Researchers have worked diligently and gained unprecedented knowledge of HIV and its interaction with the immune system. Yet, AIDS pandemic will continue for years to come. In recent years, the mathematically validated therapeutic approach is one of the most significant ways along with the biological, as well as clinical study to control the HIV for social realm of basic human rights. Its study will enable us to administer optimized level of therapies to AIDS patients and would thus be directly befitting to the society. It also provides fundamental methods and techniques for students, who are interested in epidemiological modeling, and guides junior research scientists to some frontiers at the interface of mathematical modeling and public health. It studies the dynamical behavior of the human immune system through drug ingestion under mathematical perceptive.

The book consists of mathematical modeling emphasizing HIV infection to human immune system including its responses to various available drug therapies. As mathematicians, we think that for eradication of the disease, the outline of clinical and experimental observations under the proper mathematical understanding and its application is to be more beneficial towards the society. Here we study the different drug dynamics to control the HIV disease transmission; also we study the ultimate goal of the expected time to extinction of the disease through mathematical analysis via stochastic approach. We also studied feedback mechanism in the bidirectional disease transmission dynamics which plays a significant role in combating against the infection.

Contents of the book are organized considering the process of cell biology of disease progression in HIV infection and different drug dynamics. The book helps to investigate how specific antiviral treatment can affect the immune response, that is, whether this treatment can predominantly reduce the viral load and in another sense, how it controls the disease progression in a long-term treatment of HIV infected patients. To avoid complications of the results, further analysis is performed to investigate the mathematical models with the help of optimal control theory. The effect of perfect adherence to antiretroviral therapy with respect to basic mathematical model of HIV would be introduced elaborately by impulsive differential equations. The book also studies on how we shall be able to determine the threshold value of the drug dosage and the dosing interval for which the disease can be eradicated by safe drug dosage. Delay dynamics in different variants is also discussed in this book. It is also analyzed how the delay affects the qualitative properties of the model dynamics through drug concentration during therapy. Mathematical modeling with control therapeutic approach for understanding the extinction of disease is quite significant. But under deterministic model this approach is not viable. With a view to obtain the feasibility, stochastic approach might play an escalating role in estimating the expected time to extinction of the disease in epidemiological system. In this book, we studied mathematically to find out recovery of the disease after a certain period with therapeutic control approach and later by incorporating stochastic scheme we also estimated the expected time to extinction of the disease. This book contains a mathematical-stochastic-numerical approach and analysis of different models of HIV infection with the help of control system techniques. Analytical and numerical studies and their results in treatment management would generate insights about the state of healthiness of human system, the effects of drugs including interruptions caused by HAART, IL2, DC-based immunization etc., which provides CTL-mediated control of HIV infection. This book can help to enhance the consciousness of the importance of mathematical modeling in the study of HIV/AIDS transmission and in connecting the gap between mathematical modelers in basic theoretical research, medical scientists and public health policy makers working in health research institutes. This book will appeal to undergraduate and postgraduate students and biomathematicians who are studying and working in the field of mathematical modeling on infectious diseases. Social workers who are working in the field of HIV will get prior knowledge about application of drugs to the HIV/AIDS infected patients. This book will serve as an additional textbook for graduate students and researchers in applied mathematics, health informatics, applied statistics, and qualitative public health.

Priti Kumar Roy

Acknowledgements

I convey my thanks to Shamim Ahmad, editor for Mathematical Sciences, Springer, in scripting this book on HIV. I am indebted to my scholar Dr. Amar Nath Chatterjee, for giving me the motivation and incessant effort to write a book on the subject of dynamical model of HIV. Without his active help and support, it is quite impossible task for me to complete the book.

I am grateful to Prof. Xue-Zhi Li, Department of Mathematics, Xinyang Normal University, PRC; Dr. David Greenhalgh, Department of Mathematics and Statistics, University of Strathclyde, UK; Dr. Rachel A. Norman, Department of Computing Science and Mathematics, University of Stirling, UK; Prof. R.J. Smith?, University of Ottawa, Canada; Prof. IL Hyo Jung, Pusan National University, S. Korea; Prof. Joydev Chattopadhyay, Indian Statistical Institute, Kolkata, India; Dr. Sabyasachi Bhattacharya, Indian Statistical Institute, Kolkata, India; Dr. Sumit Nandi, NIT, Kolkata, India; Dr. Sutapa Biswas Majee, NSHM Knowledge Campus, West Bengal, India from whom we enriched our knowledge regarding the biology of HIV/AIDS and their application in mathematical biology. I would like to express my gratitude to the many people who saw me through this book, to all those who provided support, talked things over, read, wrote, and offered comments, allowed me to quote their remarks and assisted in editing, proofreading and design. I am grateful to Dr. Joydeep Pal, Dr. Sonia Chowdhury, Dr. Jayanta Mondal, Mr. Abhirup Datta, Mr. Fahad Al Basir, Mr. Nikhilesh Sil, Mr. Sudip Chakraborty, Mr. Shubhankar Saha, Mr. Jahangir Chowdhury, Mr. Mithun Kumar Ghosh, Mr. Dibyendu Biswas, Ms. Priyanka Ghosh, and especially all dear colleagues of Centre for Mathematical Biology and Ecology, Department of Mathematics, Jadavpur University, for their continuous encouragement and motivation.

This book would not have been possible in its present form without the help of these people. I acknowledge the contributions of my students and those who have provided me with feedback. I would like to convey my gratitude to those who reviewed material for the book and provided valuable comments. I am sorry that I am unable to include all the topics regarding the ongoing research work, but if I had done so, the book would have run to several volumes. Finally, thanks to my family for their great support and patience to encourage me to complete this mammoth task. I humbly acknowledge all the previous researches, knowledge, and expertise in the field of HIV models which motivated us to finally write this book which would be accessible to those who are dedicated to the treatment and care of individuals affected by HIV.

I am pleased to acknowledge the financial support to research by the Government of India, Ministry of Science and Technology, Mathematical Science office No. SR/S4/MS: 558/08. I wish that you may find the book valuable and I would be eager to know about your suggestion and opinion regarding the book. I have tried my best to be error less, but it is quite an improbable task too; if you come across any error, please inform me. Finally, I would like to thank the production team of Springer for their efforts.

Priti Kumar Roy

Contents

1   **Introduction**... 1
    1.1   Historical Background of HIV/AIDS 1
    1.2   Global Scenario of HIV Pandemic....................... 2
    1.3   Disease Transmission................................. 3
    1.4   Drug Therapy Used in HIV/AIDS Treatment 6
    1.5   Aims of the Book 9
    1.6   Organization of the Book.............................. 10
    References 11

Part I   Dynamics of Immune System Against HIV

2   **Role of CTL in Restricting Virus** 19
    2.1   Suppression of CTL Responses......................... 19
        2.1.1   Equilibria.................................... 20
        2.1.2   Existence Condition 20
        2.1.3   Stability Analysis 21
        2.1.4   Numerical Simulation and Discussion 22
    2.2   Reduction of HIV Infection with Cure Rate 24
        2.2.1   Equilibria and Local Stability..................... 24
        2.2.2   Boundedness and Permanence of the System 26
        2.2.3   Global Stability of the System 27
        2.2.4   Discussion................................... 27
    2.3   Antiviral Drug Treatment along with IL-2 28
        2.3.1   Existence Condition and Stability Analysis........... 29
        2.3.2   Numerical Simulation with Discussion.............. 30
    2.4   IL-2 based Immune Therapy on T Cell 32
        2.4.1   General Analysis of the Mathematical Model 33
        2.4.2   Discussion................................... 35
    2.5   Saturation Effects for CTL-Mediated Control 35
        2.5.1   Theoretical Study of the System................... 35
        2.5.2   Existence Condition and Biological Interpretation 36

	2.5.3	Stability Analysis	36
	2.5.4	Numerical Simulation	38
	2.5.5	Discussion	38
2.6	Impact for Antigenic Stimulation on T Cell Homeostasis		39
	2.6.1	Equilibrium Points and their Stability Analysis	40
	2.6.2	Numerical Illustration and Discussion	40
References			42

3 T Cell Proliferation   43
3.1	CTL Activity through HIV Infection		43
	3.1.1	Formulation of HIV Model	43
	3.1.2	Equilibria and Local Stability	44
	3.1.3	Boundedness and Permanence of the System	46
	3.1.4	Global Stability of System	47
	3.1.5	Numerical Analysis and Discussion.	47
3.2	Effect of HAART on CTL-Mediated Immune Cells		49
	3.2.1	Equilibria.	50
	3.2.2	Stability Analysis	50
	3.2.3	Numerical Solutions of the Model Equations	53
3.3	CTL-Mediated Control of HIV Infection		53
	3.3.1	Theoretical Study of the System	54
	3.3.2	Numerical Simulation	56
References			58

4 Feedback Effect towards HIV Infection   59
4.1	Immune Cell Response to Negative Feedback Effect in HIV		59
	4.1.1	Theoretical Analysis	60
	4.1.2	Stability of the System	61
	4.1.3	Numerical Analysis.	62
	4.1.4	Discussion	64
4.2	Negative Feedback Effect in HIV Progression.		65
	4.2.1	Steady-State Analysis	66
	4.2.2	Numerical Analysis.	67
References			69

Part II Control-Based Therapeutic Approach

5 Insight of Delay Dynamics   79
5.1	Delay Effect during Long-Term HIV Infection		79
	5.1.1	Local Stability Analysis.	81
	5.1.2	Sufficient Conditions for Delay-Induced Instability	82
	5.1.3	Stability, Instability, and Bifurcation Results	83
	5.1.4	Numerical Simulations: Results and Discussions.	86
	5.1.5	Delay in Different Variants	89

5.2   Delay-Induced System in Presence of Cure Rate   91
      5.2.1   Analysis   91
      5.2.2   Numerical Simulation   94
5.3   Delay Effect during Early Stage of Infection.   95
      5.3.1   General Mathematical Model   96
      5.3.2   Numerical Simulation   100
5.4   Effect of Delay in Presence of Positive Feedback Control.   102
      5.4.1   Numerical Analysis of the Delayed System   106
5.5   Effect of Delay during Combination of Drug Therapy   107
      5.5.1   Stability Analysis of the Delay-Induced System   108
5.6   Delay-Induced System in Presence of Saturation Effect   113
References   117

6   Optimal Control Theory   119
    6.1   Optimal Control Theoretic Approach of the Implicit Model   119
          6.1.1   Numerical Simulation of the Implicit Model.   122
    6.2   The Optimal Control Problem on Chemotherapy for (3.15).   125
          6.2.1   Existence Condition of an Optimal Control   126
          6.2.2   Characterization of an Optimal Control   126
          6.2.3   Numerical Solutions of the Model Equations   128
    6.3   The Optimal Control Problem for the System (4.1)   131
          6.3.1   Numerical Analysis.   134
    6.4   The Optimal Control Problem of the System (4.5).   134
          6.4.1   Numerical Analysis.   137
    6.5   Optimization of the System (3.22).   138
          6.5.1   Numerical Simulation   140
    6.6   The Optimal Control Problem (2.11)   140
          6.6.1   Discussion   144
    6.7   Optimal Control Strategy   144
          6.7.1   Numerical Experiment of Optimal Control Strategy.   147
    6.8   The Optimal Control Problem in case of Recovery of Infected
          Cells in HIV Model.   148
          6.8.1   Numerical Simulation and Discussion   151
    References   152

7   Perfect Drug Adherence   155
    7.1   Drug Therapy with Perfect Adherence in Explicit Form   155
          7.1.1   Analysis of the Model.   156
          7.1.2   Dynamics of the Drug.   157
          7.1.3   Numerical Simulation of the Explicit Model.   160
    7.2   Enfuvirtide-IL-2 Administration for HIV-1 Treatment   162
          7.2.1   Combining T Cell Population with Virus and Drugs   162
          7.2.2   Analysis of the Model.   163
          7.2.3   Dynamics of the Drug.   164
          7.2.4   Numerical Simulation   166

    7.3   Effect of Chemokine Analog through Perfect Adherence 168
          7.3.1   Analysis of the Model 170
          7.3.2   Drug Dynamics 174
          7.3.3   Cell Count in Extreme Cases 176
          7.3.4   Numerical Simulation 178
    References 181

8 Mathematical Models in Stochastic Approach 183
    8.1   Impact for Antigenic Stimulation on T Cell Homeostasis 183
          8.1.1   Formulation of the Kolmogorov's Forward Equation 184
          8.1.2   Finding the Time to Extinction of Infected Cells 185
          8.1.3   The Distribution of the Time to Extinction 186
          8.1.4   Diffusion Approximation 188
          8.1.5   Numerical Illustration 189
          8.1.6   Discussion 191
    8.2   Expected Time to Extinction of the Disease 191
          8.2.1   Stochastic Version of the Model . . . . . . : 191
          8.2.2   The Stochastic Model Formulation 192
          8.2.3   Description of the Transition States 192
          8.2.4   Diffusion Approximation 196
          8.2.5   Expected Time to Extinction 198
          8.2.6   Numerical Simulation 199
          8.2.7   Discussion 201
    8.3   Insight of T Cell Proliferation in the Expected Time
        to Extinction 201
          8.3.1   The Deterministic Model 202
          8.3.2   The Stochastic Model Formulation 203
          8.3.3   Description of the Transition States 203
          8.3.4   Kolmogorov's Forward Equation 204
          8.3.5   Time to Extinction of Infected T Cells 205
          8.3.6   The Distribution of the Time to Extinction 207
          8.3.7   Diffusion Approximation 208
          8.3.8   The Expected Time to Extinction 210
          8.3.9   Numerical Illustration 211
          8.3.10 Discussion and Conclusion 212
    References 213

About the Author

Priti Kumar Roy is an associate professor in the Department of Mathematics, Jadavpur University, Kolkata, India. His research interests are in epidemiological issues on the chronic infectious disease such as HIV, cutaneous leishmaniasis, and filariasis. Roy also works on the neglected tropical disease like Psoriasis and has formulated robust mathematical models on the dynamics of such disease. With his mathematical expertise, he has proven the application of mathematical experimentations to provide novel solutions to mitigate the disease. He is proficient in modeling serious issues on ecological and industrial mathematics like enzyme kinematics and biodiesel production. In his publications, he has significantly contributed to boost the production of the biodiesel from *Jatropha curcas* plants and also laid down sound ecological premises to enhance the production through ecological controls that is well-applied to agro-management. He has published a significant number of publications on control therapeutic approaches and host–pathogen interactions on infectious as well as non-infectious diseases to enlighten new insights on the subjects.

With over 75 peer-reviewed research papers, Roy is a dedicated researcher in mathematical modeling and an expertise in numerical and analytical solutions to complex problems on real-life system dynamics. He has edited two books including *Insight and Control of Infectious Disease in Global Scenario* (by Intech Publishers). Three students obtained their doctoral degrees, three other scholars are waiting to obtain their doctoral degrees and eight Ph.D. students are pursuing their research works under his esteemed guidance. Roy is a renowned reviewer for various international journals of repute. He has served at several government colleges in different parts of West Bengal, India at different times. He has also delivered so many invited lectures at various foreign and Indian Universities and Institutes. He is an eminent member of several national and international societies like Biomathematical Society of India (BMSI), International Association of Engineers (IAENG), European Society of Clinical Microbiology and Infectious Diseases (ESCMID), and European Society for Mathematical and Theoretical Biology (ESMTB). He has completed six research projects as principal investigator, sponsored by various Government of India funding agencies. Roy has received the

Best Paper Award at the World Congress on Engineering 2010 held in London. He was selected as an Indian Scientist under International Collaboration/Exchange Program 2011–2012 by INSA. He was awarded with "Siksha Ratan" award in 2012.

Acronyms

APC	Antigen Presenting Cell
ART	Anti Retroviral Treatment
AZT	Azidothymidine
CTL	Cytotoxic T lymphocyte
DCs	Dendritic Cells
FO	First Order
HAART	Highly Active Anti Retroviral Therapy
HIV	Human Immunodeficiency Virus
MHC	Major Histocompatibility Complex
M–M	Michaelis–Menten
PIs	Protease Inhibitors
RTIs	Reverse Transcriptase Inhibitors
TCR	T Cell Receptor

Chapter 1
Introduction

Abstract Human immunodeficiency virus (HIV) causes severe damage of human immune system. It helps to spread the disease acquired immune deficiency syndrome (AIDS) which is a serious problem facing the human race. Thus we need serious and prompt consideration to articulate some potential treatment strategies against the disease HIV/AIDS. Our fundamental focus is to study the various mathematical models using different drugs to control the HIV/AIDS disease transmission along with available anecdotal evidences in global pharmacological practices.

Keywords AIDS/HIV · Mathematical Models · Disease

1.1 Historical Background of HIV/AIDS

In 1981, AIDS was first recognized in young healthy homosexual men in New York City, Los Angeles and San Francisco. In 1983, a cytopathic retrovirus was isolated from persons with AIDS and associated condition such as chronic lymphadenopathy. Late in 1985, serological tests had been developed and it was approved for diagnosing HIV infective people. In 1987, HIV encephalopathy, wasting syndrome, and other AIDS indicator diseases were diagnosed presumptively. Then 10 %–15 % of HIV-infected persons became reported as having AIDS. In 1993, the AIDS surveillance definition was expanded to include HIV-infected young people.

HIV had killed millions of men, women, and children from all economic classes, representing every race from countries around the world. Each day in 2003, 15000 more individuals became infected and 8000 died. The HIV epidemic in India will have a major impact on the overall spread of HIV in Asia, the Pacific, and around the world. India is second only to South Africa in terms of the overall number of people living with the disease. The first AIDS case in India was detected in 1986 and since then HIV infection has been reported in all states and union territories. The spread of HIV in India has been uneven. Although India has a low rate of infection, certain places have been more affected than others. HIV epidemics are more severe in the southern half of the country and the far northeast. The highest

© Springer Science+Business Media Singapore 2015
P.K. Roy, *Mathematical Models for Therapeutic Approaches
to Control HIV Disease Transmission*, Industrial and Applied Mathematics,
DOI 10.1007/978-981-287-852-6_1

estimated adult HIV prevalence is found in Manipur (1.40 %), followed by Andhra Pradesh (0.90 %), Mizoram (0.81 %), Nagaland (0.78 %), Karnataka (0.63 %), and Maharashtra (0.55 %). In the southern states, HIV is primarily spread through heterosexual contact. Infections in the northeast are mainly found amongst injecting drug users (IDUs) and sex workers. Unless otherwise stated, the data on this page has been taken from a 2008 report by the Government of India AIDS organization— NACO (National AIDS Control Organization) (http://www.avert.org/india-hiv-aids-statistics.htm).

1.2 Global Scenario of HIV Pandemic

HIV virus has an unexpected high mutation rate, and an infected individual often harbors many variations. The high mutation rate of HIV allows developing resistance against the drugs used to treat it. Also, some HIV-infected cells may remain in a latent stage for long period of time. Thus, the development of treatments and vaccines depends not only on knowledge of the complex life cycle of the virus, but also on understanding the complex management of the immune system operations. To control HIV, more sophisticated development of medicines and vaccines will be required.

However, poverty and politics are the major obstacle to fight against the disease. Researchers have worked diligently and gained an unprecedented knowledge of HIV and its interaction with the immune system. Yet, the AIDS pandemic will continue for years to come. Obstacles to AIDS prevention and control lie not only in the nature of the HIV virus but also in the nature of the human society worldwide. Poverty and discrimination exclude those most indeed for information and treatment. The control of HIV lies not only in biology but also in the social realm of basic human rights. AIDS is having the greatest impact in poverty ridden countries , where public health infrastructure is already strained by drug resistance malaria, tuberculosis, yellow fever, rift valley fever, and other infectious diseases. Further, the presence of HIV amplifies epidemics of such pathogens.

However, the progress against HIV is a very small effort compared to this pandemic. HIV frequency is increasing in some countries and regions, and too many new infections are still occurring. After the beginning of treatment, the number of new infections continues to reduce. However, most people are away from access to antiretroviral therapy and the demand is rising. HIV infection rates are increasing in several countries of Eastern Europe and Central Asia, which have expanding, concentrated epidemics, notably among people who inject drugs and their sexual networks. National HIV responses are too often poorly targeted to the national epidemiological situation. The HIV interventions delivered in many regions are of poor quality and do not adequately focus on vulnerable and priority risk populations in both generalized and concentrated epidemic periphery. Although variations in prevalence and epidemiological patterns within countries and regions require different priorities and interventions, all national HIV plans should incorporate service delivery to these populations in order to ensure the effectiveness of national HIV responses. In addition, those national plans need to incorporate measures to overcome structural barriers that undermine access to quality services.

1.3 Disease Transmission

HIV is transmitted principally in three ways: by sexual contact, by blood (through transfusion, blood products, or contaminated needles), and by passage from mother to child. Although homosexual contact remains a major source of HIV within United States, heterosexual transmission is the most important route of HIV spread worldwide today. Treatment of blood products and donor screening has essentially eliminated the risk of HIV from contaminated blood products in developed countries but its spread continues among intravenous drug users who share needles. In developing countries, contaminated blood and contaminated needles remain important means of infection. There are 13 %–35 % of pregnant women infected with HIV, who will pass the infection on to their babies; transmission occurs in *uteri*, as well as during birth. Breast milk from infected mothers has been shown to contain high levels of virus also. HIV is not spread by the fecal–oral route, aerosols, insects, or casual contact such as sharing household items or hugging. The risk to health works primary from direct inoculation by needle sticks and the virus cannot be spread by kissing.

Over the last several years, extensive research has been made in the understanding of the pathogenesis of HIV infection. Impressive amounts of knowledge and information have been gathered till date regarding the implications of genetic variations of immune cells. HIV pathogenesis and drugs act either by blocking the integration of viral RNA into the host CD4$^+$T cells or by inhibiting the proper cleavage of viral proteins inside an infected cell [1–3]. However, still the fundamental questions remain unanswered. On that point of view, HIV infection is very much associated with an extremely vigorous virus-specific cytotoxic T lymphocyte (CTL) response that declines disease progression [1, 2, 4–6]. When human immunodeficiency virus (HIV) invades into the body, they target the immune cells mainly CD4-positive T lymphocytes (CD4$^+$T cells, a type of white blood cells), which is the main component of immune system.

Thus, CD4$^+$T lymphocyte count is the only way to discover the disease progression monitoring during antiretroviral treatment (ART). From the premature days of infection, CD4$^+$T-lymphocyte cells have been acknowledged as most important for HIV disease progression. A healthy human adult has about 1000 CD4$^+$T cells per μl of blood [7], and when the number of CD4$^+$T cells is reduced below 200 μl, the HIV-infected patients are considered as AIDS patients [8, 9].

The infection process of the human by the HIV is a very much complicated process. When HIV invades into the body, they target the immune cells, mainly CD4$^+$T cell. HIV virus can easily hit into CD4$^+$T cells by binding process between the envelope proteins ($gp41/gp120$) on the surface of HIV with both the $CD4$ receptor and chemokine coreceptor. The HIV gains entry to the host immune cell via a series of complex steps culminating in membrane fusion of the respective cells, when the viral genome is ultimately transferred to the host cell. Entry process is initiated by binding of $gp120$, a cleaved subunit of viral surface envelope protein, Env to the $CD4^+$ receptor present on T cell surface. However, this binding will be futile unless a conformational change is induced in $gp120$ resulting in exposure of a coreceptor binding site and enabling it to bind to N-terminus of chemokine coreceptor, namely $CCR5$ and $CXCR4$, when a heterotrimeric complex of $gp120$—$CD4$-coreceptor is formed.

Basically, the chemokines are small soluble paracrine signaling molecules involved in trafficking and recruitment of leukocytes to sites of injury and inflammation. Binding to ligand transforms chemokine receptors to their signaling-active state and activate G protein coupled with receptor kinases (GRK) [10]. Chemokines and their receptors contribute significantly to disease progression in HIV-afflicted patients. Coreceptor availability and expression determine host susceptibility to infection. These coreceptors are determinants of viral tropism. Viruses can utilize either of these coreceptors as entry cofactors, when they are termed as $R5$ or $X4$ viruses. $CCR5$ tropism is essential for the onset of new infection and subsequent switching is necessary for progression of HIV to AIDS. Dual tropic or $R5X4$ viruses can utilize both of them. Binding to T cell surface receptor and coreceptor triggers membrane fusion process, when the viral fusion protein $gp41$, the second cleaved subunit of Env, undergoes reconfiguration. Thus membrane fusion is a co-operative process, where 4–6 $CCR5$ receptors, multiple $CD4$ molecules, and three to six Env trimers are involved. Receptor density and intensity of affinity between envelope protein and receptor are major determinants of viral entry efficiency [11]. After that, viral fusion is transpired in the target cell membrane and the genetic material (viral RNA) gets entry in the $CD4^+$T cell. This genetic material has a reverse transcriptase enzyme. By the reverse transcription process, the RNA genome is reverse transcribed to a DNA copy and thus the cell becomes infected. This provirus can enter into the host cell genome, where it can stay in an actively infected state or latently infected state.

The HIV is a retrovirus, whose genome is in the form of RNA and is translated into DNA during its life cycle and integrated into host cell genome. The disease (AIDS) caused by HIV is characterized by a severe impairment of the immune system and related opportunistic infections. To understand the mechanisms of HIV invading a host, let us briefly review some basic characteristics of the human immune system. The immune response of the host immune system represents a complex defense system against invading pathogens. The backbone of the human immune system is lymphocytes, which can be subdivided into B and T cells. B cells can carry antibody molecules on their cell surface and release their antibody molecules into the blood.

These antibodies then bind to the special target (such as viruses, bacteria, and protozoa) and mark it as a foreign structure for elimination by other cells (e.g., macrophages, complement system) of the immune system. T cells, unlike B cells, can only recognize protein "self" and "non-self" for the human immune system. Depending on the proteins on the surface of the cells, there are two kinds of important T cells: $CD8^+T$ cells and CD4$^+$T cells. $CD8^+T$ cells are killer cells, which have the ability to recognize and eliminate cells that are infected by special targets. However, CD4$^+$T cells are helper cells, which do not themselves kill infected cells or remove special target, but they can recognize peptides in association with major histocompatibility complex (MHC) class II molecules, and release substances that activate immune responses, which help both B cells and $CD8^+T$ cells to mount effective immune responses [12].

Dendritic cells (DCs) have a major role in HIV infection. They are the only antigen-presenting cells (APCs) that can activate naive T cells and bring out the strong T cell responses that control HIV replication. The HIV-specific DCs transport to lymphoid tissue, where they transfer HIV and stimulate HIV replication in T cells. DCs play a dual role in HIV disease. Since DCs are the APCs, they present antigen to T cells and thus T cells become infected. DCs also play a key role in intrinsic immunity through secretion of cytokines that activate natural killer (NK) cells and it produces the interferons with antiviral activity. When immature DCs come into contact with antigen, they phagocyte them and degrade their viral proteins into small pieces of antigenic peptides. After that, the immature DCs become activated to mature DCs and migrate to lymph nodes. In lymph nodes, DCs present the peptides at their cell surface using major histocompatibility (MHC) molecules and act as an APC. The antigen-presenting cell $CD8^+T$ will be activated and stimulated when they recognize and contact with the MHC class I peptide complexes [13, 14]. Matured DCs were shown to be highly efficient at stimulating CD4$^+$T cell proliferation in response to cognate peptides presented on MHC class II molecules [15]. Immature DCs convey few MHC class II molecules at their surface. Also, it is observed that DCs are infected with HIV and that infected DCs have capability to produce free virus at low levels [16].

The CD4$^+$T cells will then activate the macrophage to divide and become more aggressive. Therefore, CD4$^+$T cells are an essential part of the human immune system. However, it is unfortunate that the CD4$^+$T cell becomes the main target cell of HIV. When the HIV virus invades the CD4$^+$T cell and is inside, it starts to copy the viral RNA into DNA. The genetic machinery of this CD4$^+$T cell is manipulated to make RNA copies of the viral genome. The new viral particles are assembled on the inside of the cell membrane and leave the host cell by budding from the membrane taking the host cell membrane as new viral envelope. When thousands of virus particles bud from the host cell's surface, the CD4$^+$T cell dies, which causes the CD4$^+$T cells concentration in a host to decline.

$CD8^+$ CTLs recognize viral proteins in the form of short peptides comprising 8–11 amino acids presented in association with major histocompatibility (MHC) class I molecules on the surface of infected cells [17]. These viral peptides are derived from nascent proteins, which are cleaved by cytosolic proteases and diverted into the MHC class I antigen processing pathway throughout the process of virion synthesis [18]. Recognition of viral peptide MHC class I complexes on the surface of infected cells is a function of the T cell receptor (TCR), which in conjunction with the CD8 coreceptor mediates the translation of antigen engagement events into functional activation of the CTLs through a complex signaling cascade [19–21].

During this process, due to APC, cellular immune response activates CTL and this response/stimulation depends on the number of infected cells. Multiple mechanisms are utilized by CTLs to control viral replication [22–24]. Also during this process, due to the presence of APC, a cellular immune response activates a second type of T cell, the CTL Cell, and this response depends on the number of existing infected T cells [25]. These CTLs are only able to distinguish productively infected cells. The CTLs are the key elements of the antiretroviral immune responses. However, some virus-producing cells (infected T cell) are able to escape from this effector CTL responses, which is a determinant factor of disease progression in HIV. Also the latently infected T cell is activated by the cytokines, TCR (T cell receptor), and stimulus. Thus immunization protocols and therapeutic interventions are big challenges in AIDS research.

1.4 Drug Therapy Used in HIV/AIDS Treatment

When retroviral therapy begins in an HIV individual, the main clinical indicators of HIV-positive patients are in the follow-up of both the viral load and the CD4$^+$T cells count in blood plasma [8, 9, 26]. Also, it is to be mentioned that when therapy is started, make a portion to the immune cells to be toxic thereby introducing toxicity in the immune system of the individual. Thus qualitative aspects of the HIV-specific CTL response are to be important determinants of the efficacy of these responses in controlling viral replication. The main purpose of this study is to develop a mathematical framework that can be used to understand the various drug therapies in optimum controlled level for which it should maximize the survival time of each infected individual and minimize the number of new infections [27–29].

It has been observed clinically that patients infected with immunodeficiency virus (HIV), if treated with a combination of inhibitor drugs Lamivudine and Zidovudine, show a 10–100-fold reduction of viral load and nearly 25 % increase in the healthy CD4$^+$T cells count. Sustenance of such drug-receiving patients is observed for more than 1 year [8, 9, 30].

Recent works [8, 9, 31, 32] show that different vaccinations reduce the viral load and AIDS mortality and activate vivo-activated CTL. This vivo-activated CTL suppresses the viral replication. These vivo-activated CTLs interact with infected T cells. Due to the drug stimulation, the system generates CTL and this CTL acts against virus-producing cells (infected CD4$^+$T cells) and kills them [8, 9, 32].

HIV grows in the immune system due to its infection, but the virus still works inside the immune system. Here, the role of APC of a HIV-infected individual is thus important since it indicates precursor cytotoxic T lymphocytes to differentiate into killer T cells which are known as effectors CTL [33]. On the other hand, killer T cells instigate to destroy infected CD4$^+$T cells from where new infectious virions are born. This implies that there have been several complications in the immune system for using drugs or more precisely it can be said that the infection process is quite a complicated interaction process between different cells, viruses, and drugs [34].

Mathematical models of drug treatment dynamics suggest that the CTL response could be maintained or even increased by combining drug therapy with vaccination [35, 36]. When drug is administered in a HIV-infected individual, CTL is stimulated and it acts against the infected CD4$^+$T cells.

In recent years, the antiretroviral therapy for HIV-positive patients has largely improved. Different drug therapies are administered for different stages of the HIV-infected patients. There are more than twenty Food and Drug Administration (FDA) recommended anti-HIV drugs available. Since the primary receptor for HIV is CD4 receptor on T cell and the hallmark feature of AIDS is total impairment of host immune system with depletion of CD4$^+$T cell pool, the ultimate objective of any successful therapeutic intervention will be replenishment of CD4$^+$T count, reconstitution of immune system, and eradication of virus from the system. Administration of highly active antiretroviral therapy (HAART), comprising a combination of reverse transcriptase inhibitor (RTI) and protease inhibitor (PI), has substantially improved the well-being and life expectancy of HIV-positive patients but it has failed in certain aspects which can be overcome by adoption of complementary strategies. Most of these are divided in two categories: Reverse Transcriptase Inhibitors (RTI: AZT, ddI, ddC, D4T, 3TC, delavirdine, nevirapine, abacavir, succinate, and efavirenz) and Protease Inhibitors (PI: ritonavir, saquinavir, indinavir, and nelfinavir) [37]. The RTIs are used to prevent HIV RNA from being converted into DNA, thus blocking integration of the viral code into the target cells [37, 38]. Reverse transcriptase inhibitors prevent HIV from infecting cells or put a stop to infection of new cells. The PIs efficiently reduce the number of infectious virus particles released by an infected cell [37, 38]. Protease inhibitors prevent the production of new infectious virions by infected cells [35, 36] causing the virus to be unable to infect helper T cells. Thus, the HAART is the most effective treatment nowadays, which is mainly a combination of three or more different drugs of RTI and PI. This type of treatment is effective for preventing new infection and killing or halting the virus. HAART can achieve only partial immune reconstitution and viral reservoir of latently infected cells, which remain almost completely unaffected. Moreover, in long-term therapy, these anti-HIV agents become ineffective due to mutation in viruses.

It has also been observed that in a long-term use, these highly expensive drug therapies give results with many complications. After consuming these drugs, the patient may develop harmful side effects such as cardiovascular sensations, lactic acidosis etc. [39]. As a result, the patient may have died due to other diseases. This type of treatment for HIV does not improve the immune system to help the body to fight against any other infection. Nowadays, many clinical laboratories keep a systematic data records of patient treatment courses with respect to effectiveness and results. However, those records provide an incompatible indication as to which is better: early treatment (defined as CD4$^+$T cell counts between 200 and 500 mm^{-3} of blood) or treatment at a later stage (below 200 mm^{-3}). "Better" here is based on overall health of patient (i.e., side effects) and a preservation or amplification in the CD4$^+$T cells count [40].

To activate the immune system, cytokines play an important role. Cytokines are protein hormone and interleukin-2 (IL-2) is the main cytokine for which the immune system is activated [41]. IL-2 is a 133 amino acid protein with molecular weight of 15.5 kDa, which was described 40 years ago [42]. IL-2 is also known as T cell growth factor, as it induced T lymphocyte to enter the space of cell cycle (Tincati et al. [42]). The patients receiving IL-2 doses range in 2.5 units. The cytokine production and cellular differentiation are the effects of IL-2. Paoli and C.Tincati had studied about the effect of IL-2 on the immunological system to fight against HIV [42, 43]. IL-2 is mainly produced from CD4$^+$T cells and $CD8^+T$ cells and it acts on the same cell from which it is produced. IL-2 also activates the CD4$^+$T cells and partially $CD8^+T$ cells. In spite of CD4$^+$T cell and $CD8^+T$ cell differentiation, IL-2 also activates the latently infected cells. However, for HIV-infected patients, as the immune system become weak, IL-2 does not act accurately as its production is impaired [31]. Antiretroviral therapy can partially stimulate IL-2 production, but it does not work accurately for long period of time. Thus the immunotherapy with IL-2 is a more effective treatment to fight against the HIV infection.

In recent years, detailed insight into the involvement of chemokines and their receptors in AIDS pathogenesis has opened up a new vista in AIDS management with design of new molecules that can specifically interact with the receptors and prevent the entry of the virus particle into the cell. Furthermore, since these molecules specifically target invariant host determinants, which does not undergo mutation, they remain active even after prolonged use [44]. Chemokines can inhibit viral entry into the susceptible host cell by any one of the following mechanisms: steric hindrance due to ligand (natural or synthetic or re-engineered)–receptor interaction thereby competitively blocking viral accessor, internalization of the surface receptor through clathrin-dependent mechanism after binding to chemokine, and dimerization of chemokine receptor [44, 45].

In recent years, DC-based therapeutic vaccination has largely been developed [46, 47]. It has been observed that there are no major side effects on DC-based immunotherapy. But DC-based vaccination strategies are still at a very early stage of development for the patients with chronic infectious diseases, such as hepatitis B and C or HIV infection and no such mathematical modeling has been studied with respect to perfect drug adherence. We report a very few studies on the use of DC-based vaccines in HIV-infected patients. Lu and his coworkers studied with 18 untreated infected subjects, who were vaccinated with autologous DCs [48]. It was observed that after immunization, the median plasma viral load decreased by 80 %, eight individuals showed a decrease of more than 90 % over the period of the study (1 year), while for the other 10 the reduction was weaker and transient [49]. CD4$^+$T cell count in blood increased significantly for a short period of time (3 months), while no significant changes were observed in the $CD8^+T$ cell count [49]. The total antibodies to HIV remained unchanged after the vaccination and the neutralizing antibodies were detected at low level (1/10 titers). Garcia and his coworkers [50] carried out their study on 18 patients with chronic HIV infection undergoing HAART, who were randomized either to be vaccinated with autologous monocyte derived DCs loaded with autologous heat-inactivated HIV (12 subjects) or to represent a control group (6 subjects). After treatment (5 immunizations at 6 week intervals), HAART was interrupted and the patients were observed for at least 24 weeks to monitor safety and both the immune and clinical responses [49]. The DC-based vaccine was well accepted and no side effect was observed, with the exception of two patients who experienced mild flu-like symptoms in 24 h after immunization [49]. Here, the DC-based vaccine is in the form of live DCs. It is to be assumed that this vitro antigen-loaded DCs have similar function as the vivo antigen-loaded DCs. When this vaccine is injected to the infected individual, the viral load decreases and reaches to its minimum level.

There are several ways of drug elimination kinetics. However, when medicines are prescribed, the pharmacokinetics course of action is referred to ADME (absorption, distribution, metabolism, and excretion), which determine the drug concentration in the body [51]. Also, the effectiveness of the dosage regimen is determined by the concentration of the drug in the body. For this reason, in course of drug therapy, appropriate drug elimination kinetics makes a difference in HIV/AIDS management. There are several ways of elimination kinetics. The first-order elimination kinetics is where the amount of drug is decreasing at a rate proportional to the amount of the drug and it has been observed that most drugs are eliminated in this manner. Furthermore, if one continues to increase the amount of the drug administration, then all drugs will change from showing first-order process. In case of single substrate, biochemical reactions are represented by the M–M elimination kinetics, without observing the model's basic assumptions ("http://en.wikipedia.org/wiki/Michaelis-Mentenkinetics").

1.5 Aims of the Book

The main objective of the book is to bring in the study of the different drug dynamics to control the HIV disease transmission. Considering the process of cell biology in case of disease progression in HIV infection and different drug dynamics, the book will be organized. Our book will consist of mathematical modeling emphasizing HIV infection to human immune system including its responses to various available drug therapies. Main focus is to work on mathematical optimization and control on the models. It will enable us to administer optimized level of therapies to AIDS patients and would thus be directly benefiting to the society. The aim of this book is to provide fundamental methods and techniques for students, who are interested in epidemiological modeling, and to guide junior research scientists to some frontiers in the interface of mathematical modeling and public health.

We hope that this book can help to enhance the consciousness of the importance of mathematical modeling in the study of HIV/AIDS transmission, and in connecting the gap between mathematical modelers in basic theoretical research, medical scientists and public health policy makers working in health research institutes.

This book will contain a mathematical–statistical–computational approach of analysis with the help of control system techniques. Analytical and numerical studies and their results in treatment management would generate insights about the state of healthiness of human system, effects of drugs including interruptions caused by

HAART, IL2, DC-based immunization, etc., which provides CTL-mediated control of HIV infection. This book will appeal to undergraduate and postgraduate students and researchers who are studying and working in the field of biomathematical modeling on infectious disease. We think that a huge number of readers will be interested in the title and content of the book. The libraries of educational institutions will want the book to enrich their resources. Also, medical researchers will be benefited from this book. The social worker, those who works in the field of HIV, will get some prior knowledge about applicable drugs to the HIV-/AIDS-infected patients. This book hopefully will be an alternative and additional textbook for graduate students in applied mathematics, health informatics, applied statistics, and qualitative public health, and a useful resource for researchers in these areas.

1.6 Organization of the Book

The book is divided into two parts (I and II), which deals with CTL responses, feedback factor, delay dynamics, and controlling of disease through different drug therapies.

Part I begins with Chap. 2 which presents a brief outline of CTL responses for HIV-infected patients. In this part, we discuss different mathematical models of HIV infection to CD4$^+$T cell as a host cell including the mentioned inhibitor drug. The system responses to the drug stimulation by generating CTL and this CTLs in turn attack the actively infected CD4$^+$T cells and kill them.

In Chap. 3, we have considered the growth of CD4$^+$T cells, which is governed by a logistic equation. It is to be mentioned here that in the absence of limited population, the average specific CD4$^+$T cells' growth rate may be obtained. Our focus in this part deals specifically when the mentioned inhibitor drug is to be given to a HIV patient, what will be the dynamical behavior of the human immune system through drug stimulation by generating CTL with maximum proliferation of T cells, and that T cell population at which proliferation shuts off [8, 9, 52–54]. In this case, we have studied the mathematical model proposed by Bonhoeffer et al. [2], designed with slight modifications of the original work with introduction of two negative feedback functions (Chap. 4) justifying the inverse relationship between viral load and rate of production of uninfected cells on one hand and the decrease in strength of immune response and viral load on the other hand. Modified mathematical model of long-term viral dynamics with subsequent analysis and numerical simulation has been successfully established. The necessary conditions for the existence of three steady states with respect to feedback factor, the rate of infection, and killing rate of virus-producing cells were evaluated.

Effects of drug dynamics are examined in Part II. The part starts with Chap. 5, which discusses the delay dynamics in the process of infection between the infected and uninfected CD4$^+$T cells, through which the transmission of the disease takes place instantaneously because the natural disease transmission process requires a finite time. On this assumption, we incorporate a time lag or delay ($\tau > 0$) in the process of disease transmission in our basic mathematical model. The chapter takes the model, where we try to find out how the delay in the intercellular disease transmission affects the qualitative properties of the model dynamic drug concentration during drug therapy. We also investigate the linear stability analysis of the model. Here, we see that if the average delay is small the infected steady state is asymptotically stable and unstable, when average delay passes through its critical value. For this critical value of delay, a Hopf bifurcation occurs as the average delay passes through its critical value and a periodic solution bifurcates from the infected equilibrium. Numerical simulation confirmed the stability and bifurcation results.

In Chap. 6, we have investigated how specific antiviral treatment can affect the immune response, i.e., whether this treatment can predominantly reduce the viral load and in another sense, how it controls the disease progression in a long-term treatment of HIV-infected patients. Thus it is imperative to say that effective HAART use is oppressive for some patients and impossible to other people due to higher costs of drug and complicated course of drug therapy. To avoid such complications of the results, we have also performed a further analysis to investigate a control variable in the proposed model, where the control variable is actually the drug dose. This control affects the interaction between infected and uninfected CD4$^+$T cells and the cost function is to be maximized for the uninfected T cell, CTL population, and to be minimized for the infected T cell population and accumulated side effect.

In Chap. 7, we have also newly introduced the effect of perfect adherence to antiretroviral therapy as studied with respect to our basic mathematical model. By impulsive differential equations, we have also determined the threshold value of the drug dosage and the dosing interval for which the disease-free equilibrium remains stable. In mathematical modeling of drug dynamics, perfect adherence or partial adherence strategy can be adopted. In case of perfect adherence, the drug is given at fixed intervals for a continuous period. If the drug is administered for a time being, followed by "drug holidays," when administration is totally stopped and finally therapy is resumed, it is known as the case of partial adherence.

Finally, in Chap. 8, we have introduced the stochasticity of the existing deterministic model and try to find out time for extinction of HIV infection. Marginal distribution through quasi-stationarity added a new dimension in the epidemiological study. Here, we try to explore the expected time to extinction of the HIV infection from quasi-stationarity and observe that it depends on increasing population size.

References

1. Nowak, M.A., May, R.M.: AIDS pathogenesis: mathematical models of HIV and SIV infections. AIDS **7**, S3–S18 (1993)
2. Bonhoeffer, S., Coffin, J.M., Nowak, M.A.: Human immunodeficiency virus drug therapy and virus load. J. Virol. **71**, 3275–3278 (1997)
3. Kalamas, S.A., Goulder, P.J., Shea, A.K., Jones, N.G., Trocha, A.K., Ogg, G.S., Walke, B.D.: Levels of human immunodeficiency virus type 1-specific cytotoxic T-lymphocyte effector and

memory responses decline after suppression of viremia with highly active antiretroviral therapy. J. Virol. **73**, 6721–6728 (1999)

4. Layne, S.P., Spouge, J.L., Dembo, M.: Quantifying the infectivity of human immunodeficiency virus. Proc. Nat. Acad. Sci. USA **86**, 4644 (1989)

5. Larder, B.A., Kemp, S.D., Harrigan, P.R.: Potential mechanism for sustained antiretroviral efficiency of AZT-3TC combination therapy. Sci. **269**, 696–699 (1995)

6. Murray, J.M., Kaufmann, A.D., Kelleher, D.A.: A model of primary HIV infection. Math. Biosc. **154**, 57–85 (1998)

7. Nowak, MA., May, R.M.: Virus Dynamics, Cambridge University Press, Cambridge, UK.112 (2000)

8. Perelson, A.S., Krischner, D.E., De-Boer, R.: Dynamics of HIV infection of CD4 T cells. Math. Biosc. **114**, 81–125 (1993)

9. Perelson, A.S., Neuman, A.U., Markowitz, J.M.: Leonard, Ho, D.D.: HIV 1 dynamics in vivo: viron clearance rate, infected cell life span, and viral generation time. Science **271**, 1582–1586 (1996)

10. Lobritz, M.A., Ratcliff, A.N., Arts, E.J.: HIV-1 entry, inhibitors, and resistance. Viruses **2**, 1069–1105 (2010)

11. Robert, W., Trono, D.: The plasma membrane as a combat zone in the HIV battlefield. Genes Dev **14**, 2677–2688 (2000)

12. Xiao, D., Bossert, W.H.: An intra-host mathematical model on interaction between HIV and malaria. Bull. Math. Biol. **72**, 1892–1911 (2010)

13. Banchereau, J., Steinman, R.M.: Dendritic cells and the control of immunity. Nature **392**, 245–252 (1998)

14. Banchereau, J., Briere, F., Caux, C., Davoust, J., Lebecque, S., Liu, Y.T., Pulendran, B., Palucka, K.: Immunobiology of dendritic cells. Annu. Rev. Immunol. **18**, 767–811 (2000)

15. Steinman, R.M., Adams, J.C., Cohn, Z.A.: Identification of a novel cell type in peripheral lymphoid organs of mice identification and distribution in mouse spleen. J. Exp. Med. **141**, 804–820 (1975)

16. Donaghy, H., Gazzard, B., Gotch, F., Patterson, S.: Dysfunction and infection of freshly isolated blood myeloid and plasmacytoid dendritic cells in patients infected with HIV-1. BLOOD **101**(11), 4506–4511 (2003)

17. Townsend, A., Bodmer, H.: Antigen recognition by class I restricted T lymphocytes. Annu. Rev. Immunol. **7**, 601–624 (1989)

18. York, I., Rock, K.: Antigen processing and presentation by the class I major histocompatibility complex. Annu. Rev. Immunol. **14**, 369–396 (1996)

19. Purbhoo, M., Sewell, A.K., Klenerman, P.: Copresentation of natural HIV-1 agonist and antagonist ligands fails to induce the T cell receptor signaling cascade. Proc. Natl. Acad. Sci. USA **95**, 4527–4532 (1998)

20. Weiss, A., Littman, D.: Signal transduction by lymphocyte antigen receptors. Cell **76**, 263–274 (1994)

21. Wange, R., Samelson, L.: Complex complexes: signaling at the TCR. Immunity **5**, 197–205 (1996)

22. Berke, G.: The CTL's kiss of death. Cell **81**, 9–12 (1995)

23. Sewell, A.K., Price, D.A., Oxenius, A., Kelleher, A.D., Phillips, R.E.: Cytotoxic T lymphocyte responses to human immunodeficiency virus: control and escape. Stem Cells **18**, 233–244 (2000)

24. Tschopp, J., Hofmann, K.: Cytotoxic T cells: more weapons for new targets? Trends Microbiol. **4**, 91–94 (1996)

25. Culshaw, R.V., Ruan, S.: A delay-differentianal equation model of HIV infection of CD4$^+$T-cells. Math. Biosci. **165**, 425–444 (2000)

26. Garciá, J.A., Soto-Ramirez, L.E., Cocho, G., Govezensky, T., José, M.V.: HIV-1 dynamics at different time scales under antiretroviral therapy. J. Theor. Biol. **238**, 220–229 (2006)

27. Coffin, J.M.: HIV population dynamics in vivo: implications for genetic variation, pathogenesis, and therapy. Sci. **267**, 482–489 (1995)

28. Callaway, D.S., Perelson, A.S.: HIV-1 infection and low virul loads. Bull. Math. Biol. **64**, 29–64 (2002)
29. Culshaw, R.V., Rawn, S., Spiteri, R.J.: Optimal HIV treatment by maximising immuno response. J. Math. Biol. **48**, 545–562 (2004)
30. Zurakowski, R., Teel, A.R.: A model predictive control based scheduling method for HIV therapy. J. Theor. Biol. **238**, 368–382 (2006)
31. Kirschner, D.E., Webb, G.F.: A model of treatment strategy in the chemotherapy of AIDS. Bull. Math. Biol. **58**, 167–190 (1996)
32. Perelson, A.S., Nelson, P.W.: Mathematical analysis of HIV-1 dynamics in vivo. SIAM Theor. **41**, 3–41 (1999)
33. Altes, H.K., Wodarz, D., Jansen, V.A.A.: The dual role of CD4T helper cells in the infection dynamics of HIV and their importance for vaccination. J. Theor. Biol. **214**, 633–644 (2002)
34. Skim, H., Han, S.J., Chung, C.C., Nan, S.W., Seo, J.H.: Optimal scheduling of drug trement for HIV infection. Int. J. Control, Autom. Syst. **1**(3), 282–288 (2003)
35. Wodarz, D., Nowak, M.A.: Specific therapy regimes could lead to long-term immunological control to HIV. Proc. Natl. Acad. Sci. USA **96**(25), 14464–14469 (1999)
36. Wodarz, D., May, R.M., Nowak, M.A.: The role of antigen-independent persistence of memory cytotoxic T lymphocytes. Int. Immunol. **12**(A), 467–477 (2000)
37. Adams, B.M., Banks, H.T., Kwon, H.D., Tran, H.T.: Dynamic multidrug therapies for HIV: optimal and STI control approaches. Biosci. Eng. **1**, 223–242 (2004)
38. Kim, W.H., Chung, H.B., Chung, C.C.: Optimal switching in structured treatment interruption for HIV therapy. Asian J. Control. **8**(3), 290–296 (2006)
39. Kwon, H.D.: Optimal treatment strategies derived from a HIV model with drug-resistant mutants. Appl. Math. Comput. **188**, 1193–1204 (2007)
40. Kirschner, D., Lenhart, S., Serbin, S.: Optimal control of the chemotherapy of HIV. J. Math. Biol. **35**, 775–792 (1997)
41. Kirschner, D.E., Webb, G.F.: Immunotherapy of HIV-1 infection. J. Biol. System **6**(1), 71–83 (1998)
42. Tincati, C., Monforte, A., Marchetti, G.: Immunological mechanisims of interlukin-2(IL-2) treatment in HIV/AIDS diseases. Curr. Mol. Pharm. **2**, 40–45 (2009)
43. Paoli, D.P.: Immmunological effects of interleukin-2 therapy in human immunodeficiency virus-positive subjects. Clin. Diagn. Lab. Immunol. **8**(4), 671–677 (2001)
44. Choi, W.T., Jing, A.: Biology and clinical relevance of chemokines and chemokine receptors CXCR4 and CCR5 in human diseases. Exp. Biol. Med. **236**, 637–647 (2011)
45. Gaertnera, H., Cerinia, F., Escolaa, J.M., Kuenzia, G., Melottia, A., Offorda, R., Rossitto-Borlata, I., Nedellecc, R., Salkowitzc, J., Gorochovd, G., Mosierc, D., Hartleya, O.: Highly potent, fully recombinant anti-HIV chemokines: reengineering a low-cost microbicide. PNAS **105**(46), 17706–17711 (2008)
46. Banchereau, J., Palucka, A.K.: Dendritic cells as therapeutic vaccines against cancer. Nat. Rev. Immunol. **5**, 296–306 (2005)
47. Santini, S.M., Belardelli, F.: Advances in the use of dendritic cells and new adjuvants for the development of therapeutic vaccines. Stem Cells **21**, 495–505 (2003)
48. Lu, W., Arraes, L.C., Ferreira, W.T., Andrieu, J.M.: Therapeutic dendritic-cell vaccine for chronic HIV-1 infection. Nat. Med. **10**, 1359–1365 (2004)
49. Gessani, S., Belardelli, F.: The Biology of Dendritic Cells and HIV Infection Sandra. Springer, (2007)
50. Garcia, F., Lejeune, M., Climent, N., Gil, C., Alcami, J., Morente, V., Alos, L., Ruiz, A., Setoain, J., Fumero, E., Castro, P., Lopez, A., Cruceta, A., Piera, C., Florence, E., Pereira, A., Libois, A., Gonzalez, N., Guila, M., Caballero, M., Lomena, F., Joseph, J., Miro, J.M., Pumarola, T., Plana, M., Gatell, J.M., Gallart, T.: Therapeutic immunization with dendritic cells loaded with heat-inactivated autologous HIV-1 in patients with chronic HIV-1 infection. J. Infect Dis. **191**, 1680–1685 (2005)
51. Song, B., Lou, J., Wen, Q.: Modeling two differcnt therapy strategies for drug T-20 on HIV-1 patients. Appl. Math. Mech. **32**(4), 419–436 (2011)

52. Bao-dan, T., Yhang, Q.: Equilibrium and permanance for an autonomous competitive system
    with feedback control. Appl. Math. Sci. **50**(2), 2501–2508 (2008)
53. Wang, L., Li, M.Y.: Mathematical analysis of the global dynamics of a model for HIV infection.
    Math. Biosci. **200**, 44–57 (2006)
54. Zhou, X., Song, X., Shi, X.: A Differential equation model of HIV infection of $CD4^+T$ cells
    with cure rate. J. Math. Anal. appl. **342**, 1342–1355 (2008)

Part I
Dynamics of Immune System Against HIV

Human Immunodeficiency Virus (HIV) is a global pandemic issue which warrants urgent need of interdisciplinary studies to resolve the pathogenicity of the immune system. Pathogens are accessible to antibodies only in the blood and in the extracellular spaces. Some bacterial pathogens, parasites, and all viruses replicate inside cells where they cannot be detected by antibodies. T cells are responsible for the cell-mediated immune responses of adoptive immunity. Cell-mediated reactions depend on direct interaction between T lymphocytes and cells bearing antigen that the T cell recognize. The action of Cytotoxic T lymphocyte (CTL) is most direct. They recognize any part of the body that is infected with the viruses.

The replicating virus eventually kills the cell, releasing new virus particles. CTL can control the infection by killing the infected cell before viral replication is complete. CTLs typically express the molecule CD8 on their cell surface. $CD8^+T$ cells control the virus by either lysing the infected cell or inhibiting the HIV replication and entry into target cells. To understand the development of AIDS from HIV infection, it is important to analyze the dynamics of HIV, $CD4^+T$ cells, and $CD8^+T$ cells throughout the period of infection. $CD8^+T$ responses can be divided into (i) the lytic responses—this is also known as a direct killing process and (ii) nonlytic responses—work by inhibiting HIV replication or inhibiting viral entry into target cells known as cytokines. Nowak and Bengham [1] show that viral control depend on CTL responses and viral density. They suggested that strong CTL responses decrease the viral set point. The removal of CTLs has been shown to increase the lifespan of productively infected cells and as a result viral production increases. CTL are divided into two groups: CTL precursor (CTL_p) and CTL effector (CTL_e). CTL_p never see the antigen and do not take part in the killing of target cells. They are also called memory CTL. CTL_e differentiates from CTL_p and they carry out the killing of target cells.

From this viewpoint, HIV infection is immensely associated with an extremely vigorous virus-specific Cytotoxic T-Lymphocyte (CTL) response that declines disease progression [2–6]. When Human Immunodeficiency Virus (HIV) invades into the body, they target the immune cells, mainly $CD4$ positive T lymphocytes ($CD4^+T$ cells), which is the main component of the immune system. Over the years, extensive research had made us consolidate our understanding of the pathogenisis of HIV infection. In 1980, a direct relationship between HIV replication and the progression to AIDS was recognized [7]. In 1989, Perelson [8] developed a model for the interaction between the human immune system and HIV. Perelson and Nelson [9] extended the Perelson's model [8] and proved mathematically that the model exhibited many of the symptoms of AIDS, which were clinically observed and during the long latency period, the levels of free virus in the body reach in high level. The interaction between $CD4^+T$ cells and HIV virus was described by the simple system of differential equation proposed by Arnaout et al. [10]. This system is the most simplified model relative to the standard form.

Since free virus is short-lived relative to the infected cell [11], they have assumed viral load as proportional to the number of infected cells [6]. They made a common simplifying assumption, through which the equilibrium expression for infected cell frequency was also described. During the infection process of T cell, the thymus remains functioning. It was observed that from the existing T cell new T cells proliferate by mutation process and this T cell proliferation shuts off at its maximum population. In several mathematical models [11, 12], this proliferation process has been considered in the form of logistic growth. T cells are stimulated by antigen or mitogen, and increase its population. It is assumed that the growth of T cell is governed by a logistic equation [13]. This proliferation of T cell depends on the average degree of antigen or idiotypic network stimulation of T-cell proliferation, when it reaches its maximum threshold population the proliferation terminates [11, 9, 12]. Different approaches from the mathematical viewpoint have been developed for the drug treatment of HIV. Several pioneering research works discussed modeling on AZT (azidothymidine) treatment [12, 14, 15]. In a previous mathematical model of viral dynamics, Bonhoeffer postulated that there is no significant difference in total virus load due to drug administration. Primarily the Reverse Transcriptase Inhibitor (RTI) actually helps in recovery and restoration of uninfected healthy T-cell population [6].

HIV targets and can infect macrophages, dendritic cells (both groups express CD4 at low levels), but the major targets are CD4+T cells. It has been observed that during the non-symptomatic phase of HIV infection, the virus has a relatively low affinity toward T cells, resulting in a slow kill rate of $= CD4^+T$ cells by the immune system. Once the virus becomes lymphotropic (or T-tropic) it begins to infect $CD4^+T$ cells and the immune system is overwhelmed. When the virus acts on T cells, the $CD4^+T$ cell levels decrease to a point where the $CD4^+T$ cell population is too low to identify the antigens that could potentially be detected. The lack of antigen cover results in acquired immune deficiency syndrome (AIDS). Due to the effect of $CD4^+T$ cell depletion during AIDS various pathogens escape from T-cell recognition. As a result, various opportunistic infections occur which would be normally brought out by the helper T-cell response to avoid the immune system. Due to the impairment of immune system due to HIV infection, most infections increase in severity because the immune system's helper T cells provide a weaker contribution to a less efficient immune response. These disorders generate similar symptoms, and many of them are fatal. During the course of infection, the T-helper cell (Th) allows itself to proliferate. T cells release an intoxicating T-cell growth factor called interleukin 2 (IL-2), which acts upon itself in an autocrine fashion. Activated T cells also produce the alpha subunit of the IL-2 receptor (CD25 or IL-2R), enabling a fully functional receptor that can bind with IL-2, which activates the T cell's proliferation. The autocrine or paracrine secretion of IL-2 binding leads to motivate proliferation from existing T cells. The T cells receive both signals of activation that help to secrete IL-2, IL-4, and interferon gamma (IFN-γ). The T cells will then differentiate. T cells are normally replenished in the body. But during the course of infection the source of new T cells or the homeostatic process to control the T cell numbers is affected. The population dynamics of T cells is not well understood. New T cells are created within the body and moved to the thymus where they mature. T cells can also be created by proliferation of existing T cells in response to invasion by pathogens. In this part we represent the proliferation of T cell in the form of logistic function. We consider p as the maximum proliferation rate and T_{max} as the population density at which proliferation shuts off. There is no direct evidence of proliferation growth in the form of logistic equation. As the proliferation process is density dependent with the rate of proliferation, thus throughout the study we consider the proliferation term in the form of logistic function.

The hallmark features of disease progression of HIV infection to full-blown AIDS include T-cell hyperactivation, impairment and dysregulation of the immune system manifested by depletion of $CD4^+T$ cell and CTL exhaustion. Antigenic stimulation by HIV leads to high turnover rate of productively infected cells, disturbing T-cell homeostasis, which is re-established by supply of fresh cells from the thymus and proliferation of existing cells. Constant recruitment of $CD4^+T$ cells helps in delayed initiation of CTL activities until peak viremia is reached and its subsequent persistence leading to killing of infected cells. Loss of infected cells means lack of antigenic stimulation for CTL population resulting in a relative loss of CTL activity [10]. Thus, when viral load is considerably high, CTL-mediated killing and cytopathic effect of virus on $CD4^+T$ cells leads to exhaustion of HIV-specific immune response [16], emphasizing the existence of an indirect inverse relationship between high viral load and the density of CTL response [11]. A negative correlation may be assumed to exist between the viral load and the rate of production of uninfected target cells [17]. During low infection, therapeutic interventions using HAART-specific CTL responses gradually declines as the $CD4^+T$ cell count increases simultaneously.

An interaction is defined as positive, if activation or accumulation of a component leads to activation or accumulation of another component and negative if activation and accumulation of a component leads to inhibition or depletion of another component. If the structure of such a system is given by a certain component, which is influenced by its own activity or levels, then this component is said to regulate itself via a feedback loop. It is also known that feedback loops have been identified in a variety of regulatory systems and positive feedback control is the type of feedback when a deviation in the controlled quantity is further amplified by the control system. Since CTL production is stimulated by infected $CD4^+T$ cells, they in turn attack the actively infected $CD4^+T$ cells and kill them.

References

1. Nowak, M.A., Bangham, C.R.M.: Population dynamics of immune responses to persistent viruses. Science **272**(5258), 74–79 (1996)
2. Layne, S.P., Spouge, J.L., Dembo,M.: Quantifying the infectivity of human immunodeficiency virus. Proc. Nat. Acad. Sci. USA **86**, 4644 (1989)
3. Larder, B.A., Kemp, S.D., Harrigan, P.R.: Potential mechanism for sustained antiretroviral efficiency of AZT-3TC combination therapy. Sci. **269**, 696–699 (1995)
4. Murray, J.M., Kaufmann, A.D., Kelleher, D.A.: A model of primary HIV infection. Math. Biosc. **154**, 57–85 (1998)
5. Nowak, M.A., May, R.M.: AIDS pathogenis: mathematical models of HIV and SIV infections. AIDS **7**, S3–S18 (1993)
6. Bonhoeffer, S., Coffin, J.M., Nowak, M.A.: Human immunodeficiency virus drug therapy and virus load. J. Virol. **71**, 3275–3278 (1997)
7. Mellors, J.W., Rinaldo, J., Gupta, C., R., P., White, R.M., Todd, J.A. and Kingsley, L.A.: Prognosis in HIV-1 infection predicted by the quantity of virus in plasma. Science **272**(5265), 1167–1170 (1996)
8. Perelson, A.S.: Mathematical and Statistical Approaches to AIDS Epidemiology. Springer, Berlin. **116** (1989)
9. Perelson, A.S., Nelson, P.W.: Mathematical Analysis of HIV-1 Dynamics in Vivo. SIAM Review **41**(3–41), 122 (1999)
10. Arnaout, R.A., Nowak, M.A., Wodarz, D.: HIV-1 dynamics revisited: biphasic decay by cytotoxic T lymphocyte killing? **267**, 1347–1354 (2000)
11. Perelson, A.S., Neuman, A.U.,Markowitz, J.M., Leonard, Ho, D.D., : HIV 1 dynamics in vivo: viron clearance rate, infected cell life span, and viral generation time. Science **271**, 1582–1586 (1996)
12. Perelson, A.S., Krischner, D.E., De-Boer, R.: Dynamics of HIV infection of CD4 T cells. Math. Biosc. **114**, 81–125 (1993)
13. Roy, P.K., Chatterjee, A.N.: T-cell proliferation in a mathematical model of CTL activity through HIV-1 infection. In: Proceedings of The World Congress on Engineering 2010, WCE 2010, 30 June–2 July, London, U.K. Lecture Notes in Engineering and Computer Science, pp. 615–620 (2010)
14. Kirschner, D.E., Webb, G.F.: A model of treatment strategy in the chemotherapy of AIDS. Bull. Math. Biol. **58**, 167–190 (1996)
15. Xiao, D., Bossert, W.H.: An intra-host mathematical model on interaction between HIV and malaria. Bull. Math. Biol. **72**, 1892–1911 (2010)
16. Joly, M., Pinto, J.,M.: Role of Mathematical Modeling on the Optimal Control of HIV-1 Pathogenesis. AIChE Journal **52**, 856–884 (2006)
17. Hogue, I.B., Bajaria, S.H., Fallert, B., A., Qin S., Reinhart T.A., Kirschner, D.E.: The dual role of dendritic cells in the immune response to human immunodeficiency virus type-I infection. J Gen Virol **89**, 1–13 (2008)

Chapter 2
Role of CTL in Restricting Virus

Abstract We are interested to observe the significant role of CTL in restricting HIV. Here, we have highlighted the fact that immune system cannot produce immunity incessantly. In the chronic phase, T helper cells slowly get depleted as because these cells are exterminated by viral infection. Thus density of the HIV specific CTLs also declines. The study presents the discussion of different mathematical models of HIV infection to CD4$^+$T cells, where CD4$^+$T cells act as host cells that include the inhibitor drug. The immune system replies to the drug impetus through generation of Cytotoxic T-Lymphocyte (CTL). These CTLs attack the actively infected CD4$^+$T cells and destroy them.

Keywords Human immunodeficiency virus (HIV) · CD4$^+$T cells · Immune response · Saturation effects

2.1 Suppression of CTL Responses

Nowak and Bangham [1] considered the interaction of uninfected (x) and infected CD4$^+$T cells (y), virus population (v), and immune response (CTLs) (z) in their model.

Here we emphasize the fact that immune system cannot generate permanent immunity. In the chronic phase, T helper cells slowly decrease because these cells are killed by viral infection. As a result, density of the HIV-specific CTLs also decreases. Based on our assumptions, we moderate the model in the following form:

$$\dot{x} = \lambda - ax - \beta xv,$$
$$\dot{y} = \beta xv - dy - \alpha yz,$$
$$\dot{v} = py + \frac{gv}{b+v} - kv,$$
$$\dot{z} = \frac{syz}{1+v} - cz. \tag{2.1}$$

© Springer Science+Business Media Singapore 2015
P.K. Roy, *Mathematical Models for Therapeutic Approaches
to Control HIV Disease Transmission*, Industrial and Applied Mathematics,
DOI 10.1007/978-981-287-852-6_2

Here λ represents the natural flow rate of CD4$^+$T cells which are produced from bone marrow. These immature cells migrate to thymus and they are matured to immunocompetent T cells. The natural death rate of uninfected CD4$^+$T cell is a. The rate of infection of T cells with virus represents the parameter β and death rate of infected cell is d. The clearance rate of infected cells by CTLs is α. Free virus is produced from the infected cells at a rate p in the plasma. The term in the third equation $\frac{gv}{b+v}$ represents growth of virus from other infected cells such as macrophages and infected thymocytes. It should be noted here that the growth rate of external viral source other than T cells is g and half saturation constant of external viral source is b. Clarence rate of virus is denoted by k and the rate of CTL proliferation in response to antigen due to presence of virus is given by $\frac{syz}{1+v}$, from where s is the proliferation rate and the natural death rate of CTLs is c.

The system (2.1) needs to be analyzed with the following initial conditions: $x(0) > 0$, $y(0) > 0$, $v(0) > 0$, $z(0) > 0$.

2.1.1 Equilibria

Here, we only consider positive equilibrium points of the system and their stability. The system (2.1) with the initial conditions possesses the following positive equilibrium $E_1(x_1, 0, 0, 0)$, $E_2(x_2, y_2, v_2, 0)$, and $E^*(x^*, y^*, v^*, z^*)$, where

$$x_1 = \frac{\lambda}{a}, \, x_2 = \frac{\lambda}{a+\beta v_2}, \, y_2 = \frac{\beta\lambda v_2}{d(a+\beta v_2)},$$

$$v_2 = \frac{-(adk+dkb\beta-dg\beta-\beta\lambda p)+\sqrt{(adk+dkb\beta-dg\beta-\beta\lambda p)^2-4dk\beta(dakb-dga-\beta\lambda bp)}}{2d\beta k}, \text{ and}$$

$$x^* = \frac{\lambda}{a+\beta v^*}, \, y^* = \frac{c(1+v^*)}{s}, \, v^* = \frac{-(skb-gs-cp-cpb)+\sqrt{(skb-gs-cp-cpb)^2+4cpb(sk-cp)}}{2(sk-cp)},$$

$$z^* = \frac{\beta\lambda sv}{\alpha c(1+v)(a+\beta v)} - \frac{d}{\alpha}.$$

E_1 denotes the infection-free steady state in which the system is free of infection and immune response. Second equilibrium E_2 represents the steady state in the presence of primary infection where the immune response is not so effective. Finally, the last steady state E^* is the interior equilibrium or coexistence equilibrium point, where uninfected, infected, virus, and CTLs, i.e., the immune response are all present.

2.1.2 Existence Condition

It can be easily verified that the infected steady state E_2 exists only if $R_0 > d(1-\frac{g}{bk})$, where R_0 is the basic reproductive ratio of the form $R_0 = \frac{\lambda\beta p}{ak}$. Biologically, we can say that if the basic reproductive ratio is above some lower threshold value, then the steady state E_2 exists.

Interior equilibrium point E^* exists if (i) $x_1 - x^* > \frac{dy^*}{a}$ and (ii) $\frac{s}{c} > \frac{p}{k}$. It is known that $d > a$, and thus we can obtain $x_1 - x^* > y^*$. Biologically, we can say that the production of the uninfected CD4$^+$T cells from thymus over its life span during the infection-free steady state must be greater than the total steady state value of uninfected and infected cells after primary infection.

Second condition biologically reveals that the rate of simulation of CTL over it's life span must be greater than the rate of production of free virus from infected cells in plasma over its entire life span.

2.1.3 Stability Analysis

Here we find out the conditions for which the steady states E_1, E_2, E^* are locally asymptotically stable.

Proposition 2.1.1 *The system (2.1) is locally asymptotically stable around E_0 if $d + k > \min[\frac{g}{b}, k(\frac{R_0}{k - \frac{g}{b}} + 1)]$ is satisfied.*

Proof The eigenvalues of the above variational matrix corresponding to E_1 are $\xi_1 = -a$, $\xi_2 = -c$, and $\xi_{3,4} = \frac{-(d+k-\frac{g}{b}) \pm \sqrt{(d+k-\frac{g}{b})^2 - 4(dk - \frac{dg}{b} - \frac{p\beta\lambda}{a})}}{2}$. All the characteristic roots corresponding to E_1 will be negative if $d + k > \frac{g}{b}$ and $R_0 < d(1 - \frac{g}{bk})$. Combining these two conditions we will get expected inequality, which has been mentioned in Proposition 2.1.1. One root of the characteristic equation is positive if $R_0 > d(1 - \frac{g}{bk})$. Hence the system (2.1) is locally asymptotically stable around E_1 if the conditions are satisfied and unstable whenever E_2 exists.

Proposition 2.1.2 *The system (2.1) is locally asymptotically stable around E_2 if Routh–Hurwitz criterion is established.*

Proof The characteristic equation of system (2.1) corresponding to E_2 is given by

$$\xi^4 + B_1\xi^3 + B_2\xi^2 + B_3\xi + B_4 = 0, \tag{2.2}$$

where

$$B_1 = d + \frac{\lambda}{x_2}, \quad B_2 = \frac{d\lambda}{x_2} + p\alpha y_2 + \frac{ksy_2}{1+v_2} + \frac{bcg}{(b+v_2)^2} - \frac{bgsy_2}{(1+v_2)(b+v_2)^2} - ck,$$

$$B_3 = \frac{p\alpha\lambda y_2}{x_2} + \frac{ks\lambda y_2}{(1+v_2)x_2} + \frac{bcg\lambda}{(b+v_2)^2 x_2} + cp\beta x_2 + \frac{dksy_2}{1+v_2} + \frac{bcdg}{(b+v_2)^2}$$
$$- \frac{bgs\lambda y_2}{(1+v_2)(b+v_2)^2 x_2} - \frac{ck\lambda}{x_2} - \frac{ps\beta x_2 y_2}{1+v_2} - \frac{bdgsy_2}{(1+v_2)(b+v_2)^2} - cdk,$$

$$B_4 = cp\beta\lambda + \frac{dks\lambda y_2}{(1+v_2)x_2} + \frac{bcdg\lambda}{(b+v_2)^2 x_2} + \frac{ps\beta^2 x_2 y_2 v_2}{1+v_2} - cp\beta^2 x_2 v_2$$
$$- \frac{cd\lambda k}{x_2} - \frac{bdgs\lambda y_2}{(1+v_2)(b+v_2)^2 x_2} - \frac{ps\lambda\beta y_2}{1+v_2}.$$

Table 2.1 Parameters used in the model (2.1)

Parameter	λ	a	β	d	α	s	c	p	g	b	k
Default value	10	0.01	0.00024	0.26	0.01	1.02	0.1	11	2	14	2.4

All parameter values are taken from [2–7]

It is easy to verify that the Routh–Hurwitz criterion $B_1 > 0$, $B_4 > 0$, $\Delta = B_1 B_2 - B_3 > 0$, and $\Delta B_3 - B_1^2 B_4 > 0$ is satisfied provided $R_0 > d(1 - \frac{g}{bk})$ is true.

We use a set of realistic parameter values as in Table 2.1. When the parameter values are in Table 2.1 or nearer to that values, then it is obvious that B_4 is always positive.

Here also the last term in the above equation can be negative and it is obvious that $B_1 B_2 - B_3$ is always positive provided the parameter values are in Table 2.1 or nearer to that values.

It should be mentioned here that the last condition of R-H criterion can be verified using Mathematica.

Proposition 2.1.3 *The system (2.1) is locally asymptotically stable around E^* if Routh–Hurwitz criterion is established.*

Proof The characteristic equation of system (2.1) corresponding to E^* is given by

$$\xi^4 + A_1\xi^3 + A_2\xi^2 + A_3\xi + A_4 = 0, \tag{2.3}$$

where

$A_1 = -m_{11} - m_{22} - m_{44}$, $A_2 = m_{11}m_{22} + m_{11}m_{44} + m_{22}m_{44} - m_{23}m_{32} - m_{24}m_{42}$,
$A_3 = -m_{11}m_{22}m_{44} + m_{11}m_{23}m_{32} - m_{23}m_{34}m_{42} + m_{23}m_{32}m_{44} + m_{11}m_{24}m_{42}$
$\quad + m_{14}m_{21}m_{42}$, $A_4 = m_{11}m_{23}m_{34}m_{42} - m_{11}m_{23}m_{32}m_{44}$.

Again, $m_{11} = -\frac{\lambda}{x^*}$, $m_{14} = -\beta x^*$, $m_{21} = \beta v^*$, $m_{22} = -d - \alpha z^*$, $m_{23} = -\alpha y^*$, $m_{24} = \beta x^*$, $m_{32} = \frac{sz^*}{1+v^*}$, $m_{34} = \frac{-cz^*}{1+v^*}$, $m_{42} = p$, $m_{44} = \frac{bg}{(b+v^*)^2} - k$.

It is easy to verify that the Routh–Hurwitz criterion $A_1 > 0$, $A_4 > 0$, $\Lambda = A_1 A_2 - A_3 > 0$, and $\Lambda A_3 - A_1^2 A_4 > 0$ are satisfied provided existence condition of E^* is true. From Routh–Hurwitz criterion, A_1 is always positive from the existence conditions.

Now we verify the second R-H condition $A_4 > 0$ if $\frac{sby^*}{v^*(b+v^*)} - c > 0$, which can be verified by the numerical values of the parameter set as in Table 2.1. $A_1 A_2 - A_3 > 0$ if $0 < A_1 < a + 2\beta v^*$.

2.1.4 Numerical Simulation and Discussion

Numerically, we solve the model equation (2.1) for better understanding of the dynamical behavior of the system. Initially, we choose the default values of the

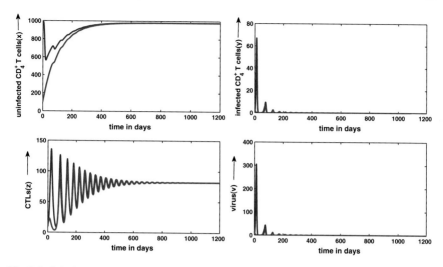

Fig. 2.1 Solution trajectories of the system (2.1) with different values of initial points, other parameters remained unchanged as in Table 2.1

parameters from their reported range in various articles. The model parameters together with their default values are given in Table 2.1. We also choose different initial values of the model variables. Figure 2.1 shows that the model variables $x(t)$, $y(t)$, and $v(t)$ oscillate initially and after some time it goes to their stable region. But $z(t)$ oscillates and after primary phase it goes to its stable region. In this case, the stability of the CTL trajectory is perturbed by virus population. It seems that the uninfected CD4$^+$T cell increases, whereas the infected cell and virus population decrease to a lower value. Figure 2.2 shows that from the initial point, trajectory initially oscillates and then converges to the interior equilibrium point.

The major difference between the model presented here and other work is that our model explores the alteration in the immune responses mediated by CTLs throughout the various stages of HIV infection starting from onset to the final manifestation in

Fig. 2.2 Solution trajectories between CTL and viral population, other parameters remained unchanged as in Table 2.1

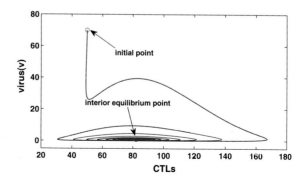

the form of AIDS. We also include in our theory anti-HIV-specific immune response. Most of the models focus on the treatment strategies using approaches specifically developed for the short-term dynamics. The main focusing point is that the HIV-specific immune response decreases due to virus population portrayed in the phase plane diagram very clearly.

2.2 Reduction of HIV Infection with Cure Rate

HIV is thought to be primarily a noncytopathic virus, primarily the virus producing cells are lost either through death, mainly immune-mediated killing or via cure that is loss of cDNA [8]. Secondly, due to antiviral therapy some infected T cells are killed by CTLs [7]. Here we suggest both cytolytic and noncytolytic mechanisms of infected cells, so that model becomes more realistic and accurate. Here we are interested to see the drug-induced changes at the steady state. Thus the free virus population is not considered as it is proportional to the virus producing cell population [7].

Since HIV is thought to be primarily a noncytopathtic virus, thus the virus producing cells are lost either by immune-mediated killing or loss of cDNA (i.e., cure). Thus we construct the model of HIV infection of CD4$^+$T cells with cure rate and the model becomes,

$$\dot{x} = \lambda - d_1 x - \beta_1 xy + \delta y,$$
$$\dot{y} = \beta_1 xy - (d_2 + \delta)y - \beta_2 yz,$$
$$\dot{z} = sy - d_3 z, \tag{2.4}$$

where λ is the constant rate of production (or supply) of immune competent T cells from the thymus, d_1 is the natural death rate of infectible cells, d_2 is the death rate of virus producing cells, and β_1 is the rate of infection of uninfected cells. Also we consider β_2 as the killing rate of virus producing cells by CTLs, s is the rate of stimulation of CTLs, and d_3 is the death rate of CTLs with the initial conditions: $x(0) > 0$, $y(0) > 0$, $z(0) > 0$. Here δ is the cure rate that is the noncytotoxic loss of virus producing cells. Thus the total disappearance of infected cells is $(d_2 + \delta)$ (Fig. 2.3).

2.2.1 Equilibria and Local Stability

The system (2.4) together with initial conditions possesses the following positive equilibria:

(i) an uninfected steady state $E_0 = (\frac{\lambda}{d_1}, 0, 0)$
(ii) an infected steady state $\bar{E} = (\bar{x}, \bar{y}, \bar{z})$, where

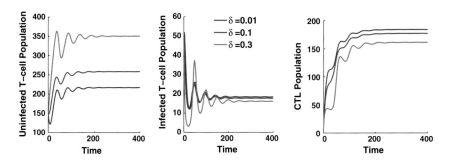

Fig. 2.3 Time series solution of the model variables for different values of δ for nondelayed system. Keeping all other parameters as in Table 2.2

Table 2.2 Parameters used in the model (2.4)

Parameter	λ	d_1	β_1	d_2	β_2	s	d_3	δ
Default value	10	0.01	0.002	0.24	0.001	0.2	0.02	0.02

All parameter values are taken from [6, 7, 9]

$$\bar{x} = \frac{(2\beta_1\delta d_3 - \beta_2 d_1 s + \beta_1 d_2 d_3) + \sqrt{(2\beta_1\delta d_3 - \beta_2 d_1 s + \beta_1 d_2 d_3)^2 - 4\beta_1^2 d_3(\delta^2 d_3 + d_2 d_3\delta - \lambda\beta_2 s)}}{2\beta_1^2 d_3},$$

$$\bar{y} = \frac{d_3(\beta_1\bar{x} - d_2 - \delta)}{s\beta_2}, \qquad \bar{z} = \frac{\beta_1\bar{x} - d_2 - \delta}{\beta_2}$$

satisfying the following inequality $\delta < \frac{2\lambda\beta_1 - d_1 d_2}{2d_1}$. Thus we get a critical value of cure rate, i.e., $\delta_{\text{crit}} = \frac{2\lambda\beta_1 - d_1 d_2}{2d_1}$. Let R_0 be the basic reproductive ratio of the system, which represents the average number of secondary infected cell caused by a single infected cell in an entirely susceptible cell population throughout its infection period. For the system (2.4) $R_0 = \frac{\lambda\beta_1 s}{d_1 d_3(d_2 + \delta)}$.

Theorem 2.2.1 *If $R_0 < 1$, $E_0 = (x_0, 0, 0)$ is locally stable, and if $R_0 > 1$, $E_0 = (x_0, 0, 0)$ is unstable.*

Theorem 2.2.2 *If (i) $R_0 > 1$, (ii) $\delta < \delta_{\text{crit}}$, and (iii) $A_1 A_2 - A_3 > 0$, then the infected steady state $\bar{E} = (\bar{x}, \bar{y}, \bar{z})$ is locally asymptotically stable.*

Proof To comment upon the existence of the local stability of the infected steady state $\bar{E}$ for $\delta < \delta_{\text{crit}}$, we consider the linearized system of (2.4) at $\bar{E}$.

Then the characteristic equation of the system becomes

$$\rho^3 + A_1\rho^2 + A_2\rho + A_3 = 0, \tag{2.5}$$

where

$A_1 = d_1 + d_2 + d_3 + \delta + \beta_1(\bar{y} - \bar{x}) + \beta_2\bar{z},$
$A_2 = d_3(d_1 + d_2 + \delta + \beta_2\bar{z} - \beta_1\bar{x}) + \beta_2 s\bar{y} + d_1(d_2 + \delta) + \beta_1(d_1\bar{z} - d_1\bar{x} + d_2\bar{y} + \beta_1\bar{y}\bar{z}),$
$A_3 = d_1 d_3(d_2 + \delta + \beta_2\bar{z} - \beta_1\bar{x}) + \bar{y}(\beta_2 s d_1 + \beta_1 d_2 d_3 + \beta_1\beta_2 d_3\bar{z} + s\beta_1\beta_2\bar{y}).$

By Routh–Hurwitz criterion, the necessary and sufficient conditions for locally asymptotic stability of the steady states are

$$A_1 > 0, \quad A_3 > 0, \quad A_1 A_2 - A_3 > 0.$$

2.2.2 Boundedness and Permanence of the System

In this section we first analyze that the system is bounded. Here we assume a positively invariant set
$\Gamma = \{(x(t), y(t), z(t) | x(t) > 0, y(t) > 0, z(t) > 0\}$. Also we assume that $x(t)$, $y(t)$, and $z(t)$ are random positive solution of the system with initial values.

Theorem 2.2.3 *The system (2.4) is bounded above for large value of $M > 0$, i.e., $x(t) \leq M$, $y(t) \leq M$, $z(t) \leq M$ for large $t \geq T$.*

Proof Let $U(t) = x(t) + y(t)$.

Now, $\dot{U} = \dot{x} + \dot{y} = \lambda - (d_1 x + d_2 y) - \beta_2 yz \leq \lambda - d_1(x + y)$. So the uninfected and infected cell population are always bounded. Thus there exists $M > 0$ such that $x(t) \leq M$, $y(t) \leq M$, $z(t) \leq M$ for large $t \geq T$. From the third equation of (2.4) it is easy to see that $z(t)$ is also bounded. Hence it is proved that system is bounded above.

Theorem 2.2.4 *The system (2.4) is bounded below for any lower value of $m > 0$, i.e., $x(t) \geq m$, $y(t) \geq m$, $z(t) \geq m$ for large $t \geq T$.*

Proof To prove the theorem we choose large $t \geq T$ such that
$\dot{y} = y(\beta_1 x - d_1 - \delta - \beta_2 z) \geq y(\beta_1 x - d_1 - \delta - \beta_2 M) \geq 0$ and
$\dot{z} = sy - d_3 z \geq sy - d_3 M \geq 0$,
for $x \geq m_1$ and $y \geq m_2$,
where $m_1 = \frac{d_1 + \delta + \beta_2 M}{\beta_1}$ and $m_2 = \frac{d_3 M}{s}$. Then $z(t)$ is also bounded below, i.e., $z(t) \geq m_3$ where $m_3 = \frac{\lambda \beta_1 - d_1(d_1 + \delta + \beta_2 M)}{\beta_1 d_3 (d_1 + \beta_2 M)}$.

Then there exists $m = \max\{m_1, m_2, m_3\}$, such that $x(t) \geq m$, $y(t) \geq m$, $z(t) \geq m$ for large $t \geq T$. Hence it is proved that the system is bounded below. Thus we can define a positive invariant set $\Gamma = \{(x(t), y(t), z(t)) | m \leq x(t) \leq M, m \leq y(t) \leq M, m \leq z(t) \leq M\}$, where each solution of the system (2.4) with positive initial value will enter in the compact region Γ and remaining finally. Summarizing the above analysis we can establish the theorem stated below.

Theorem 2.2.5 *The positive invariant solution Γ of the system (2.4) with boundedness is permanent.*

2.2.3 Global Stability of the System

Theorem 2.2.6 *For the system (2.4), $\bar{E}$ is globally asymptotically stable in Γ if $R_0 > 1$ together with the condition $A_1 A_2 - A_3 > 0$.*

Proof To prove the global stability of the system, we construct the Lyapunov function given by, $V(x(t), y(t), z(t)) = \frac{w_1}{2}(x - \bar{x})^2 + w_2(y - \bar{y} - \ln \frac{y}{\bar{y}}) + \frac{w_3}{2}(z - \bar{z})^2$. If $w_1(\beta_1 \bar{x} - \delta) = w_2 \beta_1$ and $w_2 \beta_2 = w_3 s$, then we have

$$D^+ V(x(t), y(t), z(t)) = -[w_1(d_1 + \beta_1 y)(x - \bar{x})^2 + w_3 d_3 (z - \bar{z})^2] < 0.$$

Hence we can say that the system $\bar{E}$ is globally asymptotically stable if $R_0 > 1$.

2.2.4 Discussion

In Fig. 2.4 we plot a parametric space where β_1 is plotted against s and δ. The relation between these three parameters are derived from the relation $R_0 = \frac{\lambda \beta_1 s}{d_1 d_3 (d_2 + \delta)}$. Thus the lower region of the figure represents the parametric region for which the steady state E_0 is stable and the upper region represents the region for which E_0 is unstable whereas $\bar{E}$ is asymptotically stable.

Figure 2.5 shows the phase plane portrait where β_1 is plotted against s and δ separately. In these two cases we see that β_1 decreases either as s or δ increases. Thus we can claim that if cure rate or the rate of stimulation is improved then the disease transmission rate can be controlled. Hence for nondelayed model if cure rate is improved, disease transmission would be controlled.

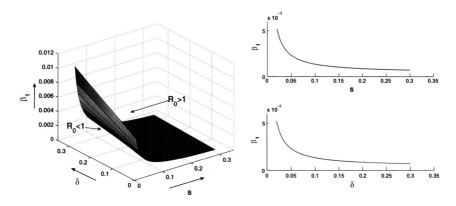

Fig. 2.4 The parametric space for δ-s-β_1. $R_0 < 1$ represents the lower surface region and $R_0 > 1$ represents the upper surface region. Keeping all other parameters as in Table 2.2. (*Top panel*) plot for the diseases transmission rate (β_1) against the rate of stimulation of CTL (s). (*Bottom panel*) plot for the diseases transmission rate (β_1) against the cure rate (δ)

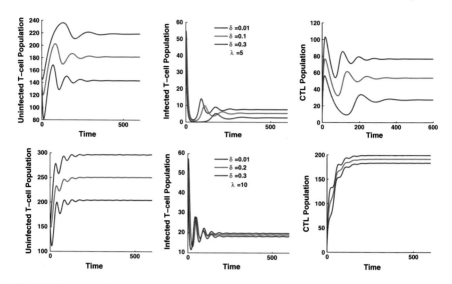

Fig. 2.5 The system behavior for different values of δ when $\lambda = 5$. Keeping all other parameters as in Table 2.2. The system behavior for different values of δ when $\lambda = 10$. Keeping all other parameters as in Table

2.3 Antiviral Drug Treatment along with IL-2

In this section, we have considered the uninfected CD4$^+$T cell population ($x(t)$), infected CD4$^+$T cell population ($y_i(t)$), latently infected cell population ($y_l(t)$), and the CTL responses ($z(t)$).

Due to immune response, cytotoxic T lymphocyte (CTL) responses at a rate s, and this response depends on the population of infected T cells. Due to this CTL response, infected T cells are killed at a rate β_2. Further in the infection process, some CD4$^+$T cells remain in the latent state. However, they are again activated and move into the infected class at a rate δ. We have chosen those infected CD4$^+$T cells that have undergone the latent stage at a rate ν. Thus, the equation of the model becomes

$$\dot{x} = \lambda - d_1 x - (1 - \eta_1 u_1)\beta_1 x y_i + \eta_2 u_2 x,$$
$$\dot{y}_i = (1 - \eta_1 u_1)\beta_1 x y_i - d_2 y_i - \beta_2 y_i z + \delta y_l,$$
$$\dot{y}_l = \nu y_i - d_3 y_l - \delta y_l,$$
$$\dot{z} = s y_i - d_4 z + \eta_3 u_2 z. \tag{2.6}$$

Here we have considered that RTI reduces the infection rate by $(1 - \eta_1 u_1)$ where η_1 is the drug efficacy and u_1 is the control input doses of the drug RTI. We have also considered the enhancement of uninfected T cells and CTL responses through IL-2 treatment, which is given by $\eta_2 u_2$ and $\eta_3 u_2$ respectively. Here u_2 denotes control input of IL-2 treatment and η_2 and η_3 are the drug efficacy of IL-2 for uninfected

CD4$^+$T cells and CTL responses respectively. Also we assume that d_3 and d_4 are the natural death rates of latently infected T cells and CTL responses.

2.3.1 Existence Condition and Stability Analysis

The system (2.6) has an infection-free equilibrium $E_0(x_0, 0, 0, 0)$ where $x_0 = \frac{\lambda}{d_1 - \eta_2 u_2}$ and E_0 exists if $d_1 > \eta_2 u_2$. Thus we can say that the infection-free equilibrium is always stable inspite of low drug dose of IL-2 compare to the death rate of uninfected CD4$^+$T cells. The another steady state is the endemic equilibrium $E^*(x^*, y_i^*, y_l^*, z^*)$ satisfying $x^* = \frac{\lambda}{d_1 + (1 - \eta_1 u_1)\beta_1 y_i^* - \eta_2 u_2}$, $\quad y_l^* = \frac{v y_i^*}{d_3 + \delta}$, $\quad z^* = \frac{s y_i^*}{d_4 - \eta_3 u_2}$, where y_i^* is a solution of

$$a_1 y_i^{*2} + a_2 y_i^* + a_3 = 0, \tag{2.7}$$

with the coefficient

$a_1 = (1 - \eta_1 u_1)s\beta_1\beta_2(d_3 + \delta),$
$a_2 = (d_3 + \delta)\{s\beta_2(d_1 - \eta_2 u_2) + d_2\beta_1(d_4 - \eta_3 u_2)(1 - \eta_1 u_1)\} - (1 - \eta_1 u_1)v\beta_1\delta(d_4 - \eta_3 u_2),$
$a_3 = (d_4 - \eta_3 u_2)\{(d_1 - \eta_2 u_2)(d_2 d_3 + d_2\delta - v\delta) - (1 - \eta_1 u_1)\lambda\beta_1(d_3 + \delta)\}.$ Now if $a_1 > 0$, $a_2 > 0$ and $a_3 < 0$, then the above equation has a unique positive root. Since $d_4 > \eta_3 u_2$ then $a_3 < 0$ for $\delta > \frac{v}{(1 - \eta_1 u_1)\beta_1\lambda}$. From the above existence condition, we can say that if the infection rate is high or the production rate of latently infected CD4$^+$T cells are very low, then with low activation rate of the latently infected CD4$^+$T cells, the endemic state may exist. The endemic state always exists for the higher production rate of uninfected CD4$^+$T cells. In a nutshell, if the production rate of uninfected T cells are high, more cells will be infected and hence latently infected cell's production will be increased. The infected steady state will be existed for the low activation rate of latently infected CD4$^+$T cells, and this activation rate depends on the immune activator IL-2. We also observe that the endemic equilibrium exists only in the absence of drug therapy. However, if the drug therapy be 100 % effective, then the endemic equilibrium does not exist.

The characteristic equation becomes,

$$(\rho + d_1 - \eta_2 u_2)(\rho + d_4 - \eta_3 u_2)[\rho^2 + A_1\rho + A_2] = 0, \tag{2.8}$$

where $A_1 = d_2 + d_3 + \delta - (1 - \eta_1 u_1)\beta_1 x_0$ and $A_2 = (d_3 + \delta)[d_2 - v\delta - (1 - \eta_1 u_1)\beta_1 x_0]$.

In the absence of RTI drug therapy (i.e., if $\eta_1 u_1 = 0$), $A_1 < 0$ and $A_2 < 0$. This means there exists at least one positive eigenvalue. Thus, we can say that in the absence of drug therapy the disease-free equilibrium is not stable, and the system moves to its endemic equilibrium. But in presence of drug therapy and if the RTI therapy is nearly 100 % effective, then $A_1 > 0$ and $A_2 > 0$ for $d_2 > v\delta$. Also all eigenvalues have negative real parts if $d_1 > \eta_2 u_2, d_4 > \eta_3 u_2$, and $d_2 > v\delta$. Hence we can say that in presence of low dosage of immune activator IL-2, if the RTI therapy is almost 100 % effective, then disease-free equilibrium will be stable.

For the endemic equilibrium the characteristic equation becomes

$$\rho^4 + B_1\rho^3 + B_2\rho^2 + B_3\rho + B_4 = 0, \qquad (2.9)$$

where

$$B_1 = [\frac{\lambda}{x^*} + d_2 + d_3 + d_4 + \delta + \beta_2 z^* - (1 - \eta_1 u_1)\beta_1 x^* - \eta_3 u_2],$$

$$B_2 = [(d_3 + \delta)(d_4 - \eta_3 u_2) + \frac{\lambda}{x^*}\{d_2 + \beta_2 z^* - (1 - \eta_1 u_1)\beta_1 x^*\} + (1 - \eta_1 u_1)^2 \beta_1{}^2$$

$$+ (1 - \eta_1 u_1)^2 \beta_1{}^2 x^* y_i{}^* + (d_3 + d_4 + \delta - \eta_3 u_2)\{\frac{\lambda}{x^*} + d_2 + \beta_2 z^*$$

$$- (1 - \eta_1 u_1)\beta_1 x^*\} - v\delta + \beta_2 s y_i{}^*],$$

$$B_3 = (d_3 + \delta)(d_4 - \eta_3 u_2)\{\frac{\lambda}{x^*} + d_2 + \beta_2 z^* - (1 - \eta_1 u_1)\beta_1 x^*\}$$

$$+ \beta_2 s y_i{}^*(\frac{\lambda}{x^*} + d_3 + \delta) - v\delta(\frac{\lambda}{x^*} + d_4 - \eta_3 u_2),$$

$$B_4 = (d_3 + \delta)(d_4 - \eta_3 u_2)[\frac{\lambda}{x^*}\{d_2 + \beta_2 z^* - (1 - \eta_1 u_1)\beta_1 x^*\} + (1 - \eta_1 u_1)^2$$

$$\times \beta_1{}^2 x^* y_i{}^*] - v\delta\frac{\lambda}{x^*}(d_4 - \eta_3 u_2) + \beta_2 s y_i{}^*\frac{\lambda}{x^*}(d_3 + \delta). \qquad (2.10)$$

Now in the absence of drug therapy (i.e., if $\eta_1 u_1 \to 0$, $\eta_2 u_2 \to 0$, and $\eta_3 u_2 \to 0$), all eigenvalues are negative if $s y_i{}^* > \frac{\beta_1}{\beta_2}$. But in the presence of drug therapy (i.e., if $\eta_1 u_1 \to 1$, $\eta_2 u_2 \to 1$, and $\eta_3 u_2 \to 1$), then at least one eigenvalue is negative. It implies that if the drug therapy is nearly or totally 100 % effective, then the endemic equilibrium will be unstable. Furthermore, in the absence of drug therapy, the endemic equilibrium is stable. Here we can conclude that in the presence of drug, the disease does not govern properly and the system is switched over to its disease's free state for sufficient drug dosage. Since HIV has other reservoirs of virus, such as follicular dendritic cells and latently infected cells [10], thus total eradication of the disease is quite impossible. However, in our model, we consider one of the virus reservoir like latently infected T cells.

2.3.2 Numerical Simulation with Discussion

We use the default values of the parameters given in Table 2.3. We have also observed that for another set of parameters, their reported range can be used and give the similar behavior.

Figure 2.6 (Left panel) shows the contour plot of the basic reproductive ratio for the system, in the absence of drug therapy as a function of λ (constant production of uninfected CD4$^+$T cell) and β_1 (the rate at which the uninfected CD4$^+$T cell becomes

Table 2.3 List of parameters for system (2.6)

Parameter	λ	d_1	d_2	d_3	d_4	β_2	δ	ν	s
Default value	10	0.03	0.3	0.03	0.2	0.001	0.1	0.5	0.2

Units are $mm^{-3}day^{-1}$ except β_2, s, and d_4 as because their units are day^{-1}

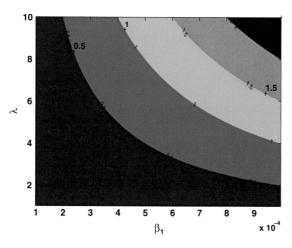

Fig. 2.6 Contour plot for basic reproductive ratio as a function of λ and β_1

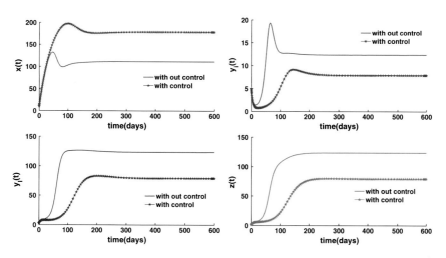

Fig. 2.7 The system behavior for without control ($u_1 = 0$, and $u_2 = 0$) and with control ($u_1 = 0.5$ and $u_2 = 0$)

infected). This figure reflects the changes of the threshold of basic reproductive ratio as β_1 and λ fluctuate. From this figure, it has been clearly observed that the disease-free equilibrium will be stable if $\lambda > 4$ and $\beta_1 < 0.0004$. But if β_1 increases and λ decreases, then the system E_0 loses its stability and thus the system moves towards the unstable region.

In Fig. 2.7, we choose $u_1 = 0.5$ and $u_2 = 0$. This means, there is no IL-2 therapy and only RTI is administered. It is clearly seen that the concentration of uninfected CD4$^+$T cells in presence of treatment, reach a higher steady state compared to the absence of treatment. Furthermore, the infected, latently infected CD4$^+$T cells, and CTL response in presence of treatment, reach a lower steady state compared to the absence of treatment.

2.4 IL-2 based Immune Therapy on T Cell

Here we develop a viral dynamical model of Wodarz and Nowak [11] by introducing IL-2 therapy in presence of HAART. The model is given below,

$$
\begin{aligned}
\dot{x}(t) &= \lambda - \beta(1 - \eta_1 u_1)x(t)y(t) - dx(t) + \gamma x(t), \\
\dot{y}(t) &= \beta(1 - \eta_1 u_1)x(t)y(t) - ay(t) - py(t)z(t), \\
\dot{w}(t) &= cx(t)y(t)w(t) - cq_1 y(t)w(t) - bw(t) + \gamma_1 w(t), \\
\dot{z}(t) &= cq_2 y(t)w(t) - hz(t),
\end{aligned}
\tag{2.11}
$$

with initial conditions: $x(0) > 0$, $y(0) > 0$, $w(0) > 0$, $z(0) > 0$.

Here x represents uninfected CD4$^+$T cells, and y, w, z are infected CD4$^+$T cells, Cytotoxic T lymphocyte precursors (CTL$_p$), CTL effector cells respectively. Here λ represents the rate of production of CD4$^+$T cells from bone marrow and these immature cells migrate to thymus and they are matured to immunocompetent T cells. The natural death rate of uninfected CD4$^+$T cells is d and β is the rate at which uninfected CD4$^+$T cells become infected. Natural death rate of infected cells is a. The clearance rate of infected cells by CTL effector is p. CTL$_p$ are assumed to proliferate in response to antigenic stimulation and then differentiate into CTL memory. The rate of proliferation of CTL$_p$ population is c and they decay at a rate b. Since the differentiation rate of precursor CTL (CTL$_p$) is not at all the same as the proliferation rate of effector CTL (CTL$_e$), we consider q_1 and q_2 as multiplicative capacity of differentiated precursor CTL and proliferated effector CTL respectively. We also assume that the removal rate of effector CTL is h.

Here we introduce IL-2 therapy in presence of HAART. We also consider that the RTI reduces the infection rate β by $(1 - \eta_1 u_1)$ where η_1 represents the drug efficacy parameter and u_1 is the control input doses of the drug RTI. By introducing interleukin protein, it enhances the growth of uninfected T cells and also in a smaller quantity, increases growth of CTL$_p$. Here γ and γ_1 are the activation rates of uninfected T cell and CTL$_p$ population respectively.

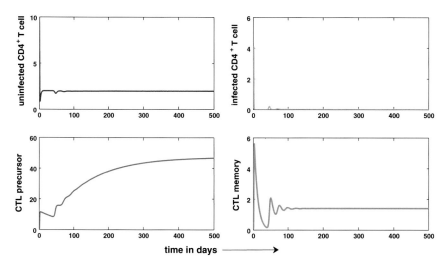

Fig. 2.8 Solution trajectory of the nondelayed system. All parameter values are taken from Table 2.4

2.4.1 General Analysis of the Mathematical Model

Figure 2.8 shows that due to introduction of cocktail drug therapy (HAART and IL-2), uninfected CD4$^+$T cell moves to its stable position. Further infected CD4$^+$T cell population moves to very lower levels, and ultimately goes towards extinction. Thus effecter CTL population attains a lower steady state, where as CTL precursor enhanced due to effect of IL-2.

2.4.1.1 Equilibria and Their Existence

The system (2.11) with the initial conditions possesses the following positive equilibrium points like $E_1(x_1, 0, 0, 0)$, $E_2(x_2, y_2, 0, 0)$, and $E^*(x^*, y^*, w^*, z^*)$, where $x_1 = \frac{\lambda}{d-\gamma}$, $x_2 = \frac{a}{\beta(1-\eta_1 u_1)}$, $y_2 = \frac{\beta(1-\eta_1 u_1)\lambda + a(\gamma - d)}{a\beta(1-\eta_1 u_1)}$, and $x^* = \frac{\lambda}{d+\beta(1-\eta_1 u_1)y^* - \gamma}$,

$$y^* = \frac{-\{(d-\gamma)cq_1 + \beta(1-\eta_1 u_1)(b-\gamma_1) - c\lambda\}}{2cq_1\beta(1-\eta_1 u_1)}$$
$$+ \frac{\sqrt{\{(d-\gamma)cq_1 + \beta(1-\eta_1 u_1)(b-\gamma_1) - c\lambda\}^2 - 4acq_1\beta(1-\eta_1 u_1)(b-\gamma_1)(d-\gamma)}}{2cq_1\beta(1-\eta_1 u_1)},$$
$$w^* = \frac{h\beta(1-\eta_1 u_1)x^* - ha}{cpq_2 y^*}, z^* = \frac{\beta(1-\eta_1 u_1)x^* - a}{p}.$$

Table 2.4 Parameters used in the model (2.11)

Parameter	λ	d	β	a	p	c	b	h	γ	γ_1	q_1	q_2
Default value	10	0.01	0.001	0.24	0.002	0.6	0.01	0.02	0.5	0.1	0.5	0.3

All parameter values are taken from [1, 2, 7, 11, 12]

During initial stages of infection when the virus enters in the system but has not yet attacked any CD4$^+$T cell, then infection-free steady state E_1 exists, if $d > \gamma$. It entails that death rate of uninfected CD4$^+$T cells is greater than the rate of production of uninfected cells under the influence of IL-2.

E_2 exists if $x_1 > x_2$, i.e., at early stages of infection when T cells have become infected but CTL response is yet to be developed and it indicates a very crucial situation. It exists when uninfected cell population at initial stage of infection is greater than steady state value of uninfected T cell population in presence of infection without any immune response.

E^* exists if the following conditions holds, (i)$\frac{\gamma - d}{\beta(1 - \eta_1 u_1)} < y^* < y_2$ and (ii)$(b - \gamma_1)(d - \gamma) < 0$.

From the above two conditions we can say that (i) if infected cell population (y^*) at coexistence equilibrium point lies between these two threshold values and (ii) product of two terms $(b - \gamma_1)(d - \gamma)$, i.e., difference between death rate and production rate of CTL$_p$ in presence of IL-2 $(b - \gamma_1)$ and difference between death rate and production rate of uninfected CD4$^+$T cell in presence of IL-2, $(d - \gamma)$ is negative.

2.4.1.2 Stability Analysis of the System

Here we study the nature of stability of the system (2.11) around different equilibrium points. From our mathematical study we have the following three propositions:

Proposition 2.4.1 *The system (2.11) is locally asymptotically stable around E_1 if the following conditions hold:*

$$(i)\ \gamma < d \quad (ii)\ \gamma_1 < b \quad (iii)\ \frac{\beta(1 - \eta_1 u_1)\lambda + a(\gamma - d)}{a\beta(1 - \eta_1 u_1)} < 0.$$

Proof The eigenvalue of the above upper triangular Jacobian matrix are

$$\xi_1 = \gamma - d, \quad \xi_2 = \frac{\beta(1 - \eta_1 u_1)\lambda}{d - \gamma} - a, \quad \xi_3 = \gamma_1 - b, \quad \xi_4 = -h.$$

All the characteristic roots corresponding to E_1 will be negative if above conditions are satisfied. Hence the system is locally asymptotically stable around E_1. Whenever E_1 is locally asymptotically stable E_2 does not exist.

Proposition 2.4.2 *The system (2.11) is locally asymptotically stable around E_2 if* $y_2 < \frac{\beta(1 - \eta_1 u_1)(b - \gamma_1)}{c(a - q_1\beta(1 - \eta_1 u_1))}$ *holds.*

Proof By the same way we can prove this proposition.

Proposition 2.4.3 *The system (2.11) is locally asymptotically stable around E^* under R-H criterion for the following parameter values in Table 2.4.*

2.4.2 Discussion

Studies first indicate the infection-free steady state or uninfected equilibrium, where CD4$^+$T cells are healthy and no infected cell population exists and there is no HIV-specific CTL immune response. The second equilibrium condition is not a very desirable one since there is no immune response even in presence of HIV antigen. The third equilibrium state or infected steady state is said to exist at certain limiting values of infected cell population in presence of HAART and IL-2. We also studied the stability of the system for different equilibrium points.

2.5 Saturation Effects for CTL-Mediated Control

In this section, we have modified exciting model of Sebastian et al. [7] considering CTL-mediated killing process including half saturation constant. Here we use $x(t)$ to represent the concentration of uninfected CD4$^+$T cells, $y(t)$ denotes the concentration of infected CD4$^+$T cells, and $z(t)$ represents density of the cytotoxic T lymphocyte (CTL) population. The model is as follows:

$$\dot{x} = \lambda - \beta xy - cx,$$
$$\dot{y} = \beta xy - by - \frac{\alpha yz}{a+y},$$
$$\dot{z} = \frac{uyz}{a+y} - dz. \tag{2.12}$$

Here λ is the constant production rate of uninfected CD4$^+$T cells which are produced from bone marrow and mature in thymus [6, 13]. Like other cells, T cells can also be removed due to their natural apoptosis, c and b are the natural death rates of uninfected and infected CD4$^+$T cells respectively, β is the rate at which infected CD4$^+$T cells infect uninfected CD4$^+$T cells. Here α is the killing rate of the infected CD4$^+$T cells by CTL, a is the half saturation constant of the killing process, d denotes the base line mortality rate of CTL, and u is the simulation rate of CTL.

The system (2.12) needs to be analyzed with the following initial conditions: $x(0) > 0$, $y(0) > 0$, $z(0) > 0$.

2.5.1 Theoretical Study of the System

2.5.1.1 Equilibria

In this section, we only consider positive equilibrium points of the system and their stability. The system (2.12) with the initial conditions possesses the following positive equilibria $E_0(x_0, 0, 0)$, $E_1(x_1, y_1, 0)$, and $E^*(x^*, y^*, z^*)$, where $x_0 = \frac{\lambda}{c}$, $x_1 = \frac{b}{\beta}$, $y_1 = \frac{\beta\lambda - bc}{b\beta}$, and $x^* = \frac{\lambda(u-d)}{ad\beta + c(u-d)}$, $y^* = \frac{ad}{u-d}$, $z^* = \frac{abcu^2 + a^2 bdu\beta + adu\beta\lambda - abcdu - au^2\beta\lambda}{\alpha(u-d)(cd - cu - ad\beta)}$.

The first equilibrium denotes the infection-free steady state in which the system is free of infection and immune response. Second equilibria E_1 represents the steady state in the presence of primary infection where the immune response is not effective. Finally, the last steady state E^* is the interior equilibria or coexistence equilibrium point.

2.5.2 Existence Condition and Biological Interpretation

The infection-free steady state E_0 always exists. The equilibrium point E_1 exists if $R_0 > 1$, as we know that $u > d$ always, where $R_0 = \frac{\lambda\beta}{bc}$ represents the basic reproductive ratio. Interior equilibrium point E^* exists if $R_0 > 1 + \frac{\beta da}{c(u-d)}$. Biologically, we can say that whenever basic reproductive ratio exceeds that upper threshold value, disease persists in the body.

2.5.3 Stability Analysis

Here we find out the conditions for which the steady states E_0, E_1, E^* are locally asymptotically stable.

Proposition 2.5.1 *The system (2.12) is locally asymptotically stable around E_0, if $R_0 < 1$.*

Proof The eigenvalues are given by,
$\xi_1 = -c$, $\xi_2 = \frac{\beta\lambda}{c} - b$, $\xi_3 = -d$. All the characteristic roots corresponding to E_0 will be negative if $\frac{\beta\lambda}{c} - b < 0$.
Hence the system (2.12) is locally asymptotically stable around E_0 if $R_0 < 1$.

Proposition 2.5.2 *The system (2.12) is locally asymptotically stable around E_1, if $1 + \frac{\beta}{c}(\frac{\lambda u}{b^2} - a) < R_0 < \min[1 + \frac{da\beta}{c(u-d)}, 1 + \frac{\beta}{c}\{\frac{a(\lambda\beta + bd)}{bu - (\lambda\beta + bd)}\}]$.*

Proof The characteristic equation of system (2.12) corresponding to E_1 is given by

$$\xi^3 - (M_{11} + M_{33})\xi^2 + (M_{11}M_{33} - M_{12}M_{21})\xi + (M_{12}M_{21}M_{33}) = 0, \quad (2.13)$$

where

$$M_{11} = -\frac{\beta\lambda}{b}, \quad M_{12} = -b, \quad M_{21} = \frac{\beta\lambda}{b} - c, \quad M_{23} = -\frac{\alpha(\lambda\beta - bc)}{ab\beta + \lambda\beta - bc},$$
$$M_{33} = \frac{u(\lambda\beta - bc)}{ab\beta + \lambda\beta - bc} - d$$

which can be expressed as

$$\xi^3 + a_1\xi^2 + a_2\xi + a_3 = 0,$$

where
$a_1 = -M_{11} - M_{33}, \quad a_2 = M_{11}M_{33} - M_{12}M_{21}, \quad a_3 = M_{12}M_{21}M_{33}.$

From Routh–Hurwitz criterion, E_1 is locally asymptotically stable if and only if $a_1 > 0$, $a_3 > 0$ and $a_1 a_2 - a_3 > 0$, where a_1 is greater than zero if $R_0 < 1 + \frac{\beta}{c}\{\frac{a(\lambda\beta+bd)}{bu-(\lambda\beta+bd)}\}$, $a_3 > 0$ if $R_0 < 1 + \frac{da\beta}{c(u-d)}$, and the last Routh–Hurwitz condition is satisfied if $R_0 > 1 + \frac{\beta}{c}(\frac{\lambda u}{b^2} - a)$. From the above three conditions, we can say that the system (2.12) is locally asymptotically stable around E_1 if $1 + \frac{\beta}{c}(\frac{\lambda u}{b^2} - a) < R_0 < \min[1 + \frac{da\beta}{c(u-d)}, 1 + \frac{\beta}{c}\{\frac{a(\lambda\beta+bd)}{bu-(\lambda\beta+bd)}\}]$ whenever it exists. Hence from the above-mentioned analysis it follows that whenever E_1 is locally asymptotically stable, E^* does not exist.

Proposition 2.5.3 *The system (2.12) is locally asymptotically stable around E^* if* $R_0 > 1 + \frac{da\beta}{c(u-d)} > \frac{dz^*\alpha(u-d)}{acu^2}$.

Proof The characteristic equation of system (2.12) is given by,

$$\xi^3 - (m_{11} + m_{22})\xi^2 + (m_{11}m_{22} - m_{23}m_{32} - m_{12}m_{21})\xi + m_{11}m_{23}m_{32} = 0,$$

where

$$m_{11} = \frac{-\beta da}{u-d} - c, m_{12} = -\frac{\beta\lambda(u-d)}{ad\beta+c(u-d)}, m_{21} = \frac{\beta da}{u-d},$$

$$m_{22} = \frac{\alpha dz^*(u-d)}{u^2 a}, m_{23} = -\frac{\alpha d}{u}, m_{32} = \frac{z^*(u-d)^2}{ua}$$

which can be expressed as,

$$\xi^3 + A_1\xi^2 + A_2\xi + A_3 = 0,$$

where
$A_1 = -m_{11} - m_{22}, \quad A_2 = m_{11}m_{22} - m_{23}m_{32} - m_{12}m_{21}, \quad A_3 = m_{11}m_{23}m_{32}.$

From Routh–Hurwitz criterion, A_1 is greater than zero if $1 + \frac{da\beta}{c(u-d)} > \frac{dz^*\alpha(u-d)}{acu^2}$. From the existence conditions of E^* we see that $A_3 > 0$ and the last R-H condition is to be satisfied if $1 + \frac{da\beta}{c(u-d)} > \frac{dz^*\alpha(u-d)}{acu^2}$. Thus, from the above analysis we can say that the system is locally asymptotically stable around E^* if $R_0 > 1 + \frac{da\beta}{c(u-d)} > \frac{dz^*\alpha(u-d)}{acu^2}$, whenever it exists.

Table 2.5 Parameters used in the model (2.12)

Parameter	λ	d	β	a	p	c	b	h	γ	γ_1	q_1	q_2
Default value	10	0.01	0.001	0.24	0.002	0.6	0.01	0.02	0.5	0.1	0.5	0.3

All parameter values are taken from [1, 2, 7, 11, 12]

2.5.4 Numerical Simulation

Numerically, we solve the model equation (2.12) for better understanding of the dynamical behavior of the system. Initially, we choose the default values of the parameters from their reported range in various articles. The model parameters together with their default values are given in Table 2.5. The initial values of the model variables are set to $x(0) = 150$, $y(0) = 30$, and $z(0) = 10$ and $x(0) = 80$, $y(0) = 25$, and $z(0) = 10$. Figure 2.9 shows that the model variables $x(t)$, $y(t)$, and $z(t)$ oscillate initially. But system moves towards its stable region as time increases. Phase portrait shows that from the different initial points, the trajectories initially oscillate and then converge to the interior point $E^*(71.76, 26.47, 10.37)$.

2.5.5 Discussion

Here we develop a population dynamical model which is an extended work of Sebastian et al. [7] representing long-term dynamics of HIV infection in response to

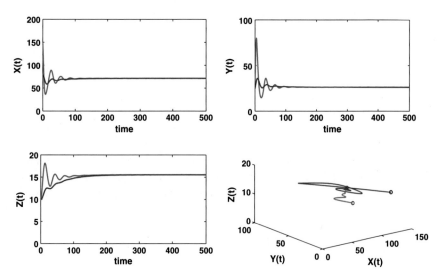

Fig. 2.9 Solution trajectories of the system (2.12) with different values of initial points, other parameter values remain unchanged as in Table 2.5

available drug therapies. We have analyzed our model analytically. In our analytical studies we have found out three equilibria with their existence conditions and also their local stability property. We have shown that if the basic reproductive ratio is greater than unity then the infection can be established in the system and whenever basic reproductive ratio crosses a certain upper threshold value, then the coexistence equilibria is locally asymptotically stable.

2.6 Impact for Antigenic Stimulation on T Cell Homeostasis

A basic mathematical model has been extensively working to describe the virus dynamics of primary HIV infection [14], which is described by a system of differential equations.

Following the above assumptions, the final set of the differential equation is given below:

$$\frac{dx}{dt} = s - dx - \beta xv + \eta_1 x,$$
$$\frac{dy}{dt} = \beta xv - [\alpha d + k_0 z]y,$$
$$\frac{dv}{dt} = py - cv,$$
$$\frac{dz}{dt} = a_E \frac{y}{y + \theta} - d_E z - \eta_2 z, \tag{2.14}$$

where x, y, v, and z represent the concentration of uninfected T cells, infected T cells, virus, and CTL responses. $s(\in R_+)$ represents the constant influx rate of target cells. The viral cytopathicity effects at a rate $\alpha(\in R_+)$, and $d(\in R_+)$ is the target cells' loss rate constant. A variable denoting free virus load in the system becomes relevant while considering short-term viral dynamics.

The process of infection to the target cells follows the law of mass action under mixing homogeneity. This means that the number of novel infection at the steady state is proportional to $x(t)y(t)$.

A per capita death rate of infected cells is considered by the parameter $\delta(\in R_+)$. The rate of contact between target cells with the virus producing infected cells represents as $\beta(\in R_+)$. By using quasi-steady state approximation, we simplify the model equations assuming that the time course of acute HIV-1 resolution is about 1 month. We obtain a model similar to the target cell limited model.

$$\frac{dx}{dt} = s - dx - \beta xv + \eta_1 x,$$

$$\frac{dy}{dt} = \beta xv - [\alpha d + k\frac{y}{y+\theta}]y,$$

$$\frac{dv}{dt} = py - cv, \tag{2.15}$$

where $k = k_0 \frac{a_E}{d_E + \eta_2}$.

2.6.1 Equilibrium Points and their Stability Analysis

The system (2.15) possesses the following equilibria: $E_1(\frac{s}{d-\eta_1}, 0, 0)$, and $E^*(x^*, y^*, v^*)$, where $v^* = \frac{py^*}{c}$, $x^* = \frac{s}{d-\eta_1+\beta'y^*}$, $\beta' = \frac{p\beta}{c}$ and y^* is the positive root of

$$\Omega_1 y^{*2} + \Omega_2 y^* + \Omega_3 = 0, \tag{2.16}$$

where $\Omega_1 = d\alpha\beta' + k\beta'$, $\Omega_2 = (d - \eta_1)(d\alpha + k) + \alpha d\theta\beta'$ and $\Omega_3 = d\alpha\theta(d - \eta_1) - s\theta\beta'$. Note that equation has a unique positive root if $\Omega_1 > 0$, $\Omega_2 > 0$, and $\Omega_3 > 0$ for which $d > \eta_1$, and y^* is positive if $\frac{s\beta'}{d\alpha(d-\eta_1)} < 1$. Now, to perform the system (2.15) which is locally asymptotically stable for different equilibrium points, we get the following conditions:
(i) $d > \eta_1$ and (ii) $\theta > \frac{s}{\alpha d}$.

2.6.2 Numerical Illustration and Discussion

In the present study, a stochastic approach has been adopted to estimate the eradication of the infection which is reflected in the expected time to extinction of the infected class. This can probably be achieved by the use of suitable drugs when we can approach at the near time to extinction of the infection stochastically. This is however the limitation which lies with the deterministic model which takes into account the death of the infected species but cannot predict the precise time to extinction. Figure 2.10 represents the dynamical behavior of the target CD4+T cells and the infected cells with variation of time. The target cell and the infected cell population display a sharp rise initially followed by a steady fall and subsequent convergence within the time range under study. The cell population attains a steady state at different values of the death rate constant α. A time delay is observed in the plot of infected cells which is an implication of the fact that infection of the target cells start after a definite period of time.

The mesh diagram in Fig. 2.11 indicates that as the viral clearance rate constant c increases, stability increases. With increase in the target cell loss rate constant (β) and the viral production (p), the zone of instability implies that infection increases with increase in target cell population. The rise in target cell population raises the

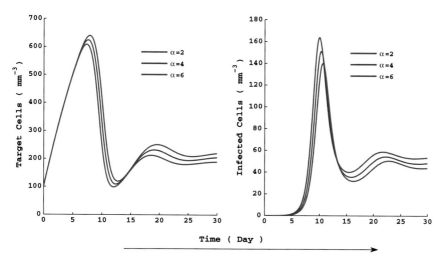

Fig. 2.10   Behavior of target CD4$^+$T-cells (x) and infected cells (y) and other parameters are as in Table 2.6

Fig. 2.11   Mesh diagram of viral clearance rate constant, viral production rate, and target cell loss rate constant with $\alpha = 2$, and other parameters are as in Table 2.6

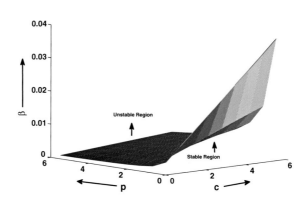

population of infected cells and it reaches an interior equilibrium point under the influence of immune capacity of the system.

This article covers the scenario of HIV infection where the CD4$^+$T cells of the body system are targeted. The CD8$^+$T effector cells spring into action and provides immunity to the system up to a certain initial stage of infection. Based on the concept of quasi-stationary distribution and diffusion approximation, the transition states are

Table 2.6   Parameters used in the model (2.14)

Parameter	s	d	p	c	β	η_1	k_0	a_E	d_E	η_2	θ
Default value	100	0.06	70	3	0.00015	0.01	0.9	2.5	1.5	0.03	1

All parameter values are taken from [14] and some are estimated outlined stochastically, which are then applied for estimating the expected time to extinction of the disease.

References

1. Nowak, M.A., Bangham, C.R.M.: Population dynamics of immune responses to persistent viruses. Science 272(5258), 74-79 (1996)
2. Culshaw, R.V., Ruan, S.: A delay-differential equation model of HIV infection of CD4$^+$T-cells. Math. Biosci. **165**, 425–444 (2000)
3. Culshaw, R.V., Rawn, S., Spiteri, R.J.: Optimal HIV treatment by maximising immuno response. J. Math. Biol **48**, 545–562 (2004)
4. Kirschner, D.E.: Using mathematics to understand HIV Immune Dynamics. Not. AMS **43**, 191–200 (1996)
5. Kirschner, D.E., Webb, G.F.: Immunotherapy of HIV-1 infection. J. Biol. System 6(1): 71–83. population dynamics. Math. Biosci. **170**, 187–198 (1998)
6. Perelson, A.S., Neuman, A.U., Markowitz, M., Leonard, J.M., Ho, D.D.: HIV 1 dynamics in vivo: viron clearance rate, infected cell life span, and viral generation time. Science 271(1582-1586), 120 (1996)
7. Bonhoeffer, S., Coffin, J.M., Nowak, M.A.: Human immunodeficiency virus drug therapy and virus load. J. Virol. **71**(3275–3278), 137 (1997)
8. Zhou, X., Song, X., Shi, X.: A differential equation model of HIV infection of $CD4^+T$ cells with cure rate. J. Math. Anal. Appl. **342**(1342–1355), 182 (2008)
9. Yang, J., Wang, X., Zhang, F.: A differential equation model of HIV infection of CD4$^+$ T cells with delay. Disc. Dyn. Nat. Soc. Article ID 903678, 16 pp. doi:10.1155/2008/903678 (2008)
10. Smith, R.J.: Explicitly accounting for antiretroviral drug uptake in theoretical HIV models predicts long-term failure of protease-only therapy. J. Theor. Biol. V 251(2), 227-237 (2008)
11. Wodarz, D., Nowak, M.A.: Specific therapy regimes could lead to long-term immunological control to HIV. Proc. Natl. Acad. Sci. USA 96(25), 14464-14469 (1999)
12. Roy, P.K., Chatterjee, A.N.: T-cell proliferation in a mathematical model of CTL activity through HIV-1 infection. In: Proceedings of The World Congress on Engineering 2010, WCE 2010, 30 June-2 July, London, U.K. Lecture Notes in Engineering and Computer Science, pp. 615-620 (2010)
13. Perelson, A.S., Krischner, D.E., De-Boer, R.: Dynamics of HIV infection of CD4+T cells. Math. Biosc. **114**(81–125), 118 (1993)
14. Burg, D., Rong, L., Neumann, A.U., Dahari, H.: Mathematical modeling of kinetics under immune control during primary HIV-1infection. J. Theor. Biol. **259**, 751–759 (2009)

Chapter 3
T Cell Proliferation

Abstract HIV infections in response to modern drug therapies have been furnished through the cell population model representing long-term dynamics of the disease. We have also considered that T cells can be created by a proliferation of existing uninfected target CD4$^+$ T cells in our human body. T cell proliferation has a noteworthy and effective role towards the disease dynamics of HIV/AIDS. Our mathematical models are based on the interactions of susceptible T cells, virus producing cells, and cytotoxic T cells that would be able to provide a complete understanding of the long-term dynamics of the system.

Keywords HIV-1 · Asymptotic stability · CD4$^+$T cells · CTL response · HAART · T cell proliferation

3.1 CTL Activity through HIV Infection

In this section, we introduce a population model representing long-term dynamics of HIV infection in response to available drug therapies. We also consider that T cells can be created by a proliferation of existing uninfected target CD4$^+$T cells in the body [1]. In this section, our mathematical model focuses on the interactions of susceptible T cells, virus producing cells and cytotoxic T cells, that would be able to provide a complete understanding of the long-term dynamics of the system.

3.1.1 Formulation of HIV Model

In this chapter, initially we consider a simple HIV model, where $x(t)$ and $y(t)$ be the uninfected and infected (virus producing cells) portions of the hosts CD4$^+$T immune cells at a time t.

It is to be assumed that T cells may be created by the proliferation of existing T cells, and the total number of T cells cannot increase unboundedly. Here, we represent the proliferation of T cells by a logistic fashion in which p is the maximum

© Springer Science+Business Media Singapore 2015

P.K. Roy, *Mathematical Models for Therapeutic Approaches*

to Control HIV Disease Transmission, Industrial and Applied Mathematics,

DOI 10.1007/978-981-287-852-6_3

proliferation rate constant, and it proliferates to a maximum level given by T_m, at which T cell proliferation shuts off. To formulate of our mathematical model, we thus consider the logistic term in the form of $px(1 - \frac{x}{T_m})$ [2, 3].

The basic virus dynamics model (as proposed by Bonhoeffer et al. [1]) with this inclusion becomes,

$$\frac{dx}{dt} = \lambda + px(1 - \frac{x}{T_m}) - d_1x - \beta xy,$$
$$\frac{dy}{dt} = \beta xy - d_2y - kyz,$$
$$\frac{dz}{dt} = sy - d_3z. \tag{3.1}$$

Here k is the killing rate of virus producing cells by CTL, s is the rate of stimulation (production) of CTL and d_3 be the baseline mortality rate of CTL. The system (3.1) needs to analyze with the following initial conditions:
$x(0) > 0$, $y(0) > 0$, $z(0) > 0$ and we denote

$$R_+^3 = \{(x, y, z) \in R^3, \ x \geq 0, \ y \geq 0, \ z \geq 0\}. \tag{3.2}$$

3.1.2 Equilibria and Local Stability

In this section, we only consider positive equilibria of the system and their stability. The system (3.1) with the initial conditions (3.2) possesses the following positive equilibriums $E_0(x_0, 0, 0)$ and $E^*(x^*, y^*, z^*)$ where,

$$x_0 = \frac{T_m}{2p}[(p - d_1) + \sqrt{(p - d_1)^2 + 4\frac{p\lambda}{T_m}}], \tag{3.3}$$

and

$$x^* = \frac{(\frac{d_2d_3\beta}{sk} + p - d_1) + \sqrt{(\frac{d_2d_3\beta}{sk} + p - d_1)^2 + 4\lambda(\frac{d_3\beta^2}{sk} + \frac{p}{T_m})}}{2(\frac{d_3\beta^2}{sk} + \frac{p}{T_m})},$$
$$y^* = \frac{d_3}{sk}(\beta x^* - d_2), \quad z^* = \frac{\beta x^* - d_2}{k}. \tag{3.4}$$

It satisfies the following inequalities

$$p > (d_1 - \frac{d_2d_3\beta}{sk}) \quad \text{and} \quad x^* > \frac{d_2}{\beta}. \tag{3.5}$$

To study the stability of the following equilibrium let us introduce the basic reproductive ratio R_0 defined as $R_0 = \frac{x_0}{x^*}$ of the system (3.1).

If $R_0 > 1$, then the positive equilibrium $E^*(x^*, y^*, z^*)$ exists.

For the equilibrium $E_0(x_0, 0, 0)$, the characteristic equation becomes,

$$(\rho - p + d_1 + \frac{2px_0}{T_m})(\rho + d_2 - \beta x_0)(\rho + d_3) = 0, \tag{3.6}$$

whose eigen values are

$$\rho_1 = p - d_1 - \frac{2px_0}{T_m} = -(\frac{px_0}{T_m} + \frac{\lambda}{x_0}) < 0, \quad \rho_2 = -d_3 < 0,$$

$$\text{and} \quad \rho_3 = \begin{cases} \beta x_0 - d_2 < 0, & \text{if } R_0 < 1, \\ \beta x_0 - d_2 > 0, & \text{if } R_0 > 1. \end{cases} \tag{3.7}$$

Thus, we can establish the proposition.

Proposition 3.1.1 *If $R_0 < 1$, then E_0 is locally asymptotically stable and if $R_0 > 1$,
then E_0 is unstable.*

Now for the equilibrium $E^*(x^*, y^*, z^*)$ the characteristic equation is as follows,

$$\rho^3 + A\rho^2 + B\rho + C = 0, \tag{3.8}$$

where

$$(i) \quad A = \frac{px^*}{T_m} + \frac{\lambda}{x^*} + d_3 > 0,$$

$$(ii) \quad B = d_3(\frac{px^*}{T_m} + \frac{\lambda}{x^*}) + sky^* + \beta^2 x^* y^* > 0,$$

$$(iii) \quad C = sky^*(\frac{px^*}{T_m} + \frac{\lambda}{x^*}) + d_3\beta^2 x^* y^* > 0. \tag{3.9}$$

From the Routh–Hurwitz criterion, the necessary and sufficient condition for local
asymptotic stability of the steady state is

$$AB - C > 0, \tag{3.10}$$

and the system locally asymptotically stable around the positive interior equilibrium if,

$$(i) \quad x^* > \frac{d_2}{\beta}, \quad (ii) \quad d_1 > \beta, \quad (iii) \quad p > \frac{(d_1 d_2 - \lambda)\beta^2 T_m}{d_2(\beta T_m - d_2)},$$

$$(iv) \quad \frac{d_3\beta^2}{sk} + \frac{p}{T_m} > 0, \quad (v) \quad p > d_1 - \frac{d_2 d_3 \beta}{sk},$$

$$(vi)\ \ \beta > \frac{-d_2 p T_m + \sqrt{(d_2 p T_m)^2 + 4(d_2^2 - \frac{1}{T_m})[T_m(\lambda - d_1 d_2) + \frac{d_2}{sk}]p}}{2[T_m(\lambda - d_1 d_2) + \frac{d_2}{sk}]}. \quad (3.11)$$

Then, we can establish the proposition.

Proposition 3.1.2 *If (i)$R_0 > 1$ and (ii) $AB - C > 0$, then the positive equilibrium $E^*(x^*, y^*, z^*)$ is locally asymptotically stable.*

3.1.3 Boundedness and Permanence of the System

To discuss the permanence of the system (3.1) we assume $R_3^+ = \{(x(t), y(t), z(t)) | x(t) > 0, y(t) > 0, z(t) > 0\}$ a positively invariant set. Also we assume that $x(t)$, $y(t)$, and $z(t)$ are random positive solution of the system with initial values. To prove the permanence of the system, we first prove the boundedness by using some theorem given below.

Proposition 3.1.3 *There is $M > 0$ such that for any positive $x(t)$, $y(t)$, and $z(t)$ of the system (3.1), $x(t) \leq M$, $y(t) \leq M$, and $z(t) \leq M$ for large t.*

Proof Let $L(t) = x(t) + y(t)$.

According to the system (3.1), we can find $\frac{dL}{dt} \leq -hL + M_0$, where $M_0 = \frac{p^2 T_m + 4\lambda p}{4p}$. Then, there exists $M_1 > 0$ depending upon the parameter of the system (3.1) such that $L(t) \leq M_1$ for large $t \geq T$. Hence $x(t)$ and $y(t)$ are bounded above. From the third equation of the system (3.1), $z(t)$ is also bounded above. Now, for large $t \geq T$, there exists M such that $x(t) \leq M$, $y(t) \leq M$ and $z(t) \leq M$. Therefore it is proven that the system is bounded above.

Proposition 3.1.4 *The system (3.1) satisfies the initial conditions (3.2) and there exists m such that $x(t) \geq m$, $y(t) \geq m$, $z(t) \geq m$ for large $t \geq T$.*

Proof To prove the following lemma, we choose large $t \geq T$ such that

$$\frac{dy}{dt} = y(\beta x - d_2 - kz) \geq y(\beta x - d_2 - kM) \geq 0 \text{ and}$$

$$\frac{dz}{dt} = sy - d_3 z \geq sy - d_3 M \geq 0, \quad \text{for } x \geq m_1 \text{ and } y \geq m_2,$$

$$\text{where } m_1 = \frac{d_2 + kM}{\beta} \quad \text{and} \quad m_2 = \frac{M d_3}{s}. \quad (3.12)$$

Here $z(t)$ is also bounded below, i.e., $z(t) \geq m_3$, where $m_3 = \frac{s}{d_3 \beta}[\frac{\lambda \beta}{d_2 + Mk} + p(1 - \frac{d_2 + Mk}{T_m \beta}) - d_2]$.

Thus, there exists $m = \max(m_1, m_2, m_3)$ such that $x(t) \geq m$, $y(t) \geq m$, and $z(t) \geq m$ for large $t \geq T$. Hence, it is proven that the system is bounded below.

Therefore we can define,
$D = \{x(t), y(t), z(t) | m \le x(t) \le M, \ m \le y(t) \le M, \ m \le z(t) \le M\}$, where D is ultimately bounded set of the system (3.1) where each solution of the system with positive initial value will be entered the compact region D and remain it finally. So we have the following persistence proposition.

Proposition 3.1.5 *The positive invariant solution of the system (3.1) with boundedness is permanent.*

3.1.4 Global Stability of System

If the system (3.1) with initial conditions (3.2) and the inequalities (3.11), then the equilibrium $E^*(x^*, y^*, z^*)$ is local asymptotic stable. We construct the Liapunov function as,

$$V(x, y, z) = w_1(x - x^* - x^* \ln \frac{x}{x^*}) + w_2(y - y^* - y^* \ln \frac{y}{y^*}) + \frac{w_3}{2}(z - z^*)^2.$$

$$(3.13)$$

Calculating the upper right derivative $V(x, y, z)$ along the system (3.1) ($w_1 = w_2 = sw$ and $w_3 = kw$) we have,

$$D^+V(t) = -[w_1(\frac{\lambda}{xx^*} + \frac{p}{T_m})(x - x^*)^2 + w_3 d_3(z - z^*)^2] < 0. \quad (3.14)$$

Therefore, according to the Liapunov function we have the following proposition,

Proposition 3.1.6 *If the system (3.1) satisfies (3.2) and the inequality (3.11), then the equilibrium $E^*(x^*, y^*, z^*)$ is globally asymptotically stable.*

3.1.5 Numerical Analysis and Discussion

Theoretical analysis of the model is done to explore equilibria and their stability of the solutions. It has been proven that the positive invariant solution of the system with boundedness is permanent. Our analytical solutions also reveal that the system moves to the global stable regime. However, for the physical realization of the time evolution of different populations with varying model parameters, we consider numerical solutions of the set of equations (3.1). This enables us to visualize the dynamical behaviors of variables x, y, and z. Initially, we choose the default values of the parameter from their reported range of various articles [1, 3–5]. Numerical solutions of the model equations (3.1) are done with the basic model parameters set to their standard values as in Table 3.1. At $t = 0$, values of model variables are considered as $x(0) = 50$, $y(0) = 50$, and $z(0) = 2$ and the units are mm^{-3}.

In the first panel (left side) of Fig. 3.1 we find an interior equilibrium point $E^*(591.19, 7.4903, 188.62)$ which is asymptotically stable with the default parameter values given in the Table 3.1. Here, we are interested to see the effect of the dynamical system due to the change in parameter values of the model. In Fig. 3.1 we plot a time series of the model variables for different values of proliferation rate (p). From this figure it is clearly observed that the uninfected T cell population rises proportionately with the rate of proliferation of uninfected T cells (p) and as p increases with its reported range, the system moves towards its stable region more quickly. We also see that if p is small, then there is an oscillation in early stage, but for the large value of p the system moves towards its equilibrium point with a short period of time.

In support of our analytical and numerical results, we plot a mesh diagram Fig. 3.2 with the parameters p, k, and β. Trajectories show that as p and k increase, then β reaches to the threshold value 0.000158. Also it is clearly manifested that the disease can be controlled. Moreover, a shift of the system to the globally stable regime carries along with the assurance of eradication of toxic T cells from the immune system naturally, as represented in our analysis. It is to be concluded from analytical and numerical findings that if the proliferation rate of targeted T cells and CTL responses

Table 3.1 Parameters used in the model (3.1)

Parameter	λ	p	T_m	d_1	β	d_2	k	s	d_3
Default value	10	0.03	1500	0.01	0.002	0.24	0.001	0.2	0.02

All parameter values are taken from [1, 3–5] and some are estimated

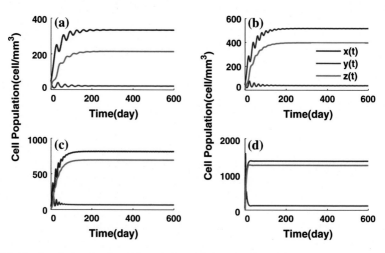

Fig. 3.1 The Population density of the uninfected, infected cells and CTL responses converges to their equilibrium. Time series solution of the model variables for different values of the proliferation rate. Various model parameters are as in Table 3.1. **a** p = 0.03, **b** p = 0.108, **c** p = 0.3, **d** p = 3

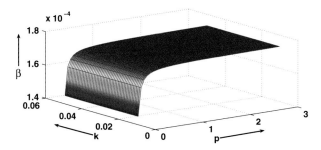

Fig. 3.2 Graphical representation of the stability criteria of E^* in p, k, and β parametric space within the range $0.01 \leq s \leq 0.05$ and $0.03 \leq p \leq 3$

is improved by the proper drug dosage, then infection rate can be controlled and as a result the virus producing cell population reduces. Thus by maintaining the proper drug therapy we can improvise the system for better outcome.

3.2 Effect of HAART on CTL-Mediated Immune Cells

To study the immune responses against HIV in presence of drug therapy, we have considered previous mathematical model [6] with slight modification. It is to be noted here that, our immune system induces CTL responses through the interactions with dendritic cells, and we think about that CTL response include both CTL effector and memory CTL [7]. Again, in the absence of immune impairment effect, proliferation of CTL depends upon the host population and viral replication together with the initial conditions [7, 8]. By the above-mentioned assumptions, we can write down the following set of differential equations under HIV induced immunological system as

$$
\begin{aligned}
\dot{x} &= \lambda + px\left(1 - \frac{x}{T_m}\right) - d_1 x - \beta_1 xy, \\
\dot{y} &= \beta_1 xy - d_2 y - \beta_2 yz_d - \beta_3 yz_e, \\
\dot{z}_d &= sy - d_3 z_d, \\
\dot{z}_e &= cxyz_e - d_4 z_e.
\end{aligned}
\tag{3.15}
$$

Here, the proliferation of effector CTLs are described by $cxyz_e$, where c is the proliferation rate, d_4 is the rate of decay, and β_3 is the killing rate of virus producing cell by z_e. The above system has to be analyzed with the following initial conditions:

$$\{x(0) > 0, \ y(0) > 0, \ z_d(0) > 0, \ \text{and} \ z_e(0) > 0\}.$$

We observe that the right-hand side of the above function is a smooth function of the variable (x, y, z_d and z_e) and the parameters with their non negative conditions. So local existence and uniqueness properties holds in R_+^4.

3.2.1 Equilibria

We find that the system has following equilibria $E_0(x_0, 0, 0, 0)$, $E_1(x_1, y_1, 0, 0)$, $E'(x', y', z_d', 0)$, and $E^*(x^*, y^*, z_d^*, z_e^*)$ where

$$x_0 = \frac{T_m}{2p}[(p - d_1) + \sqrt{(p - d_1)^2 + \frac{4p\lambda}{T_m}}], \quad x_1 = \frac{d_2}{\beta_1}, \quad y_1 = \lambda + \frac{d_2(p - d_1)}{\beta_1} - \frac{d_1^2 p}{\beta_1^2 T_m},$$

$$x' = \frac{(\frac{\beta_1 d_2 d_3}{\beta_2 s} - d_1 + p) + \sqrt{(\frac{\beta_1 d_2 d_3}{\beta_2 s} - d_1 + p)^2 + 4\lambda(\frac{p}{T_m} + \frac{d_1 \beta_1^2}{s\beta_2})}}{2(\frac{p}{T_m} + \frac{d_1 \beta_1^2}{s\beta_2})}, \quad y' = \frac{d_3(\beta_1 x_3 - d_2)}{s\beta_2}, \quad z_d' = \frac{\beta_1 x_3 - d_2}{\beta_2}, \text{ and}$$

$$x^* = \frac{T_m}{2pc}[c(p - d_1) + \sqrt{c^2(p - d_1)^2 + \frac{4pc}{T_m}(c\lambda - \beta_1 d_4)}], \quad y^* = \frac{d_4}{cx^*}, \quad z_d^* = \frac{sd_4}{d_3 cx^*},$$

$$z_e^* = \frac{d_3 cx^*(\beta_1 x^* - d_2) - s\beta_2 d_4}{d_3 c\beta_3 x^*}.$$

(i) The first equilibrium E_0 represents the uninfected equilibrium with the minimum level of healthy $CD4^+T$ cells and there exists no infected cell as well as CTL response (both drugs induced and effector CTL).

(ii) The second equilibrium E_1 represents the presence of uninfected and infected $CD4^+T$ cells and there exists no immune response. This system is very harmful to HIV-infected patients.

(iii) The third equilibrium E' represents the presence of uninfected and infected T cells together with drug activated CTL. In this case there exists no effector CTL.

(iv) The interior equilibrium E^* represents the presence of uninfected T-cells, infected T-cells, drug-induced CTL and effector CTL.

Remark 3.2.1 Existence condition for E_0: $p > d_1$, for E_1: $p > \frac{\beta_1 T_m(d_1 d_2 - \lambda \beta_1)}{d_2(\beta_1 T_m - d_2)}$, for E': $p > (d_1 - \frac{d_2 d_3 \beta_1}{s\beta_2})$ and for E^*: $p > \frac{\beta_1 d_1 T_m}{\beta_1 T_m - d_2}$, if $c > \frac{\beta_1 d_4}{\lambda}$.

It is observed that E_1 arises for E_0 if $p = \frac{\beta_1 T_m(d_1 d_2 - \lambda \beta_1)}{d_2(\beta_1 T_m - d_2)}$ and exists for all $p > \frac{\beta_1 T_m(d_1 d_2 - \lambda \beta_1)}{d_2(\beta_1 T_m - d_2)}$. We also observe that the equilibrium E' arises for E_1 if $p = (d_1 - \frac{d_2 d_3 \beta_1}{s\beta_2})$ and exists for $p > (d_1 - \frac{d_2 d_3 \beta_1}{s\beta_2})$. For $p > \frac{\beta_1 d_1 T_m}{\beta_1 T_m - d_2}$, and $c > \frac{\beta_1 d_4}{\lambda}$, E^* becomes stable and E' does not exist.

3.2.2 Stability Analysis

In this section, we establish stability results for steady state E_0, E_1, E', E^* respectively.

Proposition 3.2.1 *The uninfected equilibrium E_0 of the system (3.15) is stable for* $x_0 < \frac{d_1}{\beta_1}$, $p > d_1$ *and for* $|p| < \frac{\alpha_2' + \sqrt{\alpha_2'^2 - 4\alpha_1' \alpha_3'}}{2\alpha_1'}$, *where*, $\alpha_1' = 1 - (T_m - \frac{2d_2}{\beta_1})^2$, $\alpha_2' = 2d_1 - \frac{4\lambda}{T_m} + \frac{4d_1 d_2 T_m}{\beta_1} - 2T_m^2 d_1$, *and* $\alpha_3' = d_1^2(1 - T_m^2)$.

Proof The characteristic equation of the linearized system of (3.15), corresponding to E_0, is given by

$$\{\rho - p(1 - \frac{2x_0}{T_m}) + d_1\}(\rho - \beta_1 x_0 + d_2)(\rho + d_3)(\rho + d_4) = 0. \quad (3.16)$$

The eigen values are $p(1 - \frac{2x_0}{T_m}) - d_1, \beta_1 x_0 - d_2, -d_3,$ and $-d_4$.

Since one eigen value is negative (i.e, $\rho = -d_4 < 0$), the local stability of E_0 demands the negative real parts of all roots of the equation, which gives $x_0 < \frac{d_2}{\beta_1}$, and $|p| < \frac{\alpha_2' + \sqrt{\alpha'^2 - 4\alpha_1 \alpha_3}}{2\alpha_1}$, where $\alpha_1' = 1 - (T_m - \frac{2d_2}{\beta_1})^2, \alpha_2' = 2d_1 - \frac{4\lambda}{T_m} + \frac{4d_1 d_2 T_m}{\beta_1} - 2T_m^2 d_1,$ and $\alpha_3' = d_1^2(1 - T_m^2)$.

For the above condition E_0 is stable and when this inequality is reversed, E_0 loses stability, and the infected steady state becomes a locally asymptotically stable spiral point. Thus, we can say that if the proliferation rate is restricted, and the uninfected T cell population is less than $\frac{d_2}{\beta_1}$, then the system is locally stable around E_0.

Proposition 3.2.2 *The system (3.15) is locally asymptotically stable around E_1 if* $p > \max\{(d_1 + d_2 + d_3) - \frac{s\beta_2}{2\beta_1}, -d_2 + \sqrt{d_2^2 + (2d_1 + 3d_2 + 2d_3)\beta_1}\}$.

Proof The characteristic equation of the linearized system of (3.15), corresponding to E_1, is given by

$$(\rho + d_4 - cx_1 y_1)(\rho^3 + A_1 \rho^2 + A_2 \rho + A_3) = 0, \quad (3.17)$$

where $A_1 = d_1 + d_2 + d_3 + \beta_1 y_1 - p + x_1(\frac{2p}{T_m} - \beta_1)$,
$A_2 = d_2 d_3 + s\beta_2 y_1 - d_3 \beta_1 x_1 + (d_2 + d_3 - \beta_1 x_1)\{d_1 + \beta_1 y_1 - p(1 - \frac{2x_1}{T_m})\}$,
$A_3 = d_3 \beta_1^2 x_1 y_1 + (d_2 d_3 + s\beta_2 y_1 - d_3 \beta_1 x_1)\{d_1 + \beta_1 y_1 - p(1 - \frac{2x_1}{T_m})\}$. Since one eigenvalue is negative, $\rho = -(d_4 - cx_1 y_1)$, thus the system is locally stable around E_1 if $\rho^3 + A_1 \rho^2 + A_2 \rho + A_3 = 0$ has all eigen values with negative real part. Now it is obvious that according to Routh–Hurwith criteria E_1 is asymptotically stable if $p < d_1, p > \max\{(d_1 + d_2 + d_3) - \frac{s\beta_2}{2\beta_1}, -d_2 + \sqrt{d_2^2 + (2d_1 + 3d_2 + 2d_3)\beta_1}\}$. Thus, we can say that if the proliferation rate is below the death rate of uninfected T cells then in the presence of infection the system is stable around E_1.

Proposition 3.2.3 *The system (3.15) is locally asymptotically stable around E' if*

(i) $\frac{d_2 \beta_1^2 + \sqrt{d_2^2 \beta_1^4 + 4sd_2 \beta_1^3 \beta_2}}{2\beta^3} < x' < \min\{\frac{d_3}{\beta_1}, \frac{s\beta_2 + d_2 \beta_1}{\beta_1^2}\}$

(ii) $s\beta_2 < \frac{d_3^2 - d_2 d_3 \beta_1}{d_2 \beta_1}$.

Proof The characteristic equation is

$$(\rho - d_4 + cx'y')(\rho^3 + B_1 \rho^2 + B_2 \rho + B_3) = 0, \quad (3.18)$$

where

$$B_1 = d_2 + d_3 + \beta_2 z_d' + \frac{\lambda}{x'} + \frac{px'}{T_m} - \beta_1 x',$$

$$B_2 = s\beta_2 y' + d_2 d_3 + d_2 \beta_2 z_d' - d_3 \beta_1 x' + \beta_1{}^2 x' y' + (\frac{\lambda}{x'} + \frac{px'}{T_m})(d_2 + d_3 + \beta_2 z_d' - \beta_1 x'),$$

$$B_3 = (\frac{\lambda}{x'} + \frac{px'}{T_m})(s\beta_2 y' + d_2 d_3 + d_2 \beta_2 z_d' - d_3 \beta_1 x') + d_3 \beta_1{}^2 x' y'. \tag{3.19}$$

From the above characteristic equation we can see one eigenvalue is negative, i.e., $\rho = -(d_4 - cx' y')$.

If the equation $\rho^3 + B_1 \rho^2 + B_2 \rho + B_3 = 0$ has all roots with negative real parts and satisfies Routh–Hurwith criterion then we can easily say that the system E' is locally stable. It can be easily seen that $B_1 > 0$, $B_2 > 0$, $B_3 > 0$, and $B_1 B_2 - B_3 > 0$ if

$$\frac{d_1 \beta_1{}^2 + \sqrt{d_1 \beta_1{}^2 + 4 s d_1 \beta_1{}^3 \beta_2}}{2\beta^3} < x' < \min\{\frac{d_3}{\beta_1}, \frac{s\beta_2 + d_2 \beta_1}{\beta_1{}^2}\},$$

$$s\beta_2 < \frac{d_3{}^2 - d_2 d_3 \beta_1}{d_2 \beta_1}. \tag{3.20}$$

Therefore, the steady state E' is locally asymptotically stable if the above conditions are satisfied.

Proposition 3.2.4 *The interior equilibrium E^* is locally asymptotically stable if*
$p > \max\{\frac{\beta_1 T_m}{2}, \frac{\beta_1(d_1 + d_3)}{\beta_1 - \frac{4pd_2}{T_m}}, \frac{d_1 \beta_1}{d_3 \beta_1 + \frac{2d_2 d_3}{T_m}}, \frac{cd_3 \beta_1{}^2}{2sd_4 \beta_2(1 + \frac{1}{T_m})}\}$, $p < \frac{sd_4{}^2 \beta_2 T_m}{2pd_3 c^2 x^{*2}}$,
$c > \frac{\beta_1(\beta_1 d_4 + d_3 - d_3 \beta_1)}{d_3(d_2 + \beta_2)}$, $\beta_3 d_4 < 1$, *and* $\frac{d_4(d_2 + \beta_2)}{s\beta_2 + \beta_1} < x^* < \frac{d_3}{d_3 + \beta_1}$.

Proof The characteristic equation of the linearized system of (3.15), corresponding to E^*, is given by

$$\rho^4 + C_1 \rho^3 + C_2 \rho^2 + C_3 \rho + C_4 = 0, \tag{3.21}$$

where $C_1 = \frac{\lambda}{x^*} + \frac{px^*}{T_m} + d_3$, $C_2 = d_3(\frac{\lambda}{x^*} + \frac{px^*}{T_m}) + \beta_1{}^2 x^* y^* + s\beta_2 y^* + c\beta_3 y^{*2} z_e{}^*$,
$C_3 = (\frac{\lambda}{x^*} + \frac{px^*}{T_m})(s\beta_2 y^* + c\beta_3 y^{*2} z_e{}^*) + (d_3 + \beta_1)c\beta_3 x^* z_e{}^* + d_3 \beta_1 x^* y^*$,
$C_4 = cd_3 \beta_3 y^{*2} z_e{}^*(\frac{\lambda}{x^*} + \frac{px^*}{T_m} + \beta_1)$, and $\Delta = C_1 C_2 - C_3$.

From the above characteristic equation it is obvious that $C_1 > 0$, $C_4 > 0$, and $\Delta > 0$. Therefore, using Routh–Hurwitz criterion, we find that all roots of the characteristic equation for E^* will have negative real parts if they satisfy the conditions given below

$$p > \max\{\frac{\beta_1 T_m}{2}, \frac{\beta_1(d_1 + d_3)}{\beta_1 - \frac{4pd_2}{T_m}}, \frac{d_1 \beta_1}{d_3 \beta_1 + \frac{2d_2 d_3}{T_m}}, \frac{cd_3 \beta_1{}^2}{2sd_4 \beta_2(1 + \frac{1}{T_m})}\}, \quad p < \frac{sd_4{}^2 \beta_2 T_m}{2pd_3 c^2 x^{*2}},$$

$c > \frac{\beta_1(\beta_1 d_4 + d_3 - d_3 \beta_1)}{d_3(d_2 + \beta_2)}$, $\beta_3 d_4 < 1$, and $\frac{d_4(d_2 + \beta_2)}{s\beta_2 + \beta_1} < x^* < \frac{d_3}{d_3 + \beta_1}$. Thus, the system is asymptotically stable around the interior equilibrium E^*, if it satisfies the above conditions.

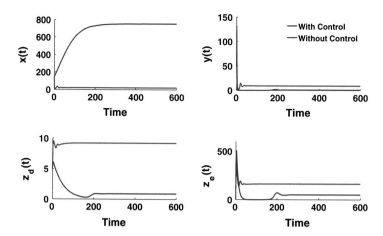

Fig. 3.3 Time series solution for the cases non treated (without control) and treated (with control) keeping $u = 0.01$ and all other parameters as in Table 3.1

3.2.3 Numerical Solutions of the Model Equations

In this section, we illustrate without control and with control model numerically. In the numerical simulation we assume $x(0) = 100$, $y(t) = 50$, $z_d(t) = 2$, $z_e(t) = 5$ and the units are mm^{-3}.

In Fig. 3.3 the number of uninfected T cell decreases rapidly with a short period and then reaches to its stable region. However, if the drug control is used, then uninfected T cell population increases and reach to its adequate level. In this figure, we also see that the number of infected T cell decreases, but it does not extinct. Here, we also see that both the Lymphocyte cell population (Drug-induced CTL and Effector CTL) decreases very fast for treated patients compared to the uncontrolled problem.

3.3 CTL-Mediated Control of HIV Infection

The positive feedback mechanism plays a vital role in CTL responses and the new T cells are produced from the existing T cells. Taking these two important factors into consideration to the existing model of Sebastian et al. [1] takes the following form:

$$\dot{x} = \lambda + rx\left(1 - \frac{x}{c}\right) - \beta xy - dx,$$
$$\dot{y} - \beta xy - ay - \rho yz,$$
$$\dot{z} = \frac{ky^n}{1 + ky^n} - bz,$$

$$(3.22)$$

under the initial conditions $x(0) > 0, y(0) > 0, z(0) > 0$. Here $x(t)$, $y(t)$, and $z(t)$ represent infectible CD4$^+$T cells, virus producing CD4$^+$T cells, and cytotoxic T lymphocyte (CTLs) responses respectively. The healthy activated CD4$^+$T cells are produced from a source such as the thymus, at a constant rate λ and d is the natural death rate of infectible CD4$^+$T cells. In the presence of HIV, T cells become infected. The number of new infections at the steady state is proportional to $x(t)y(t)$, where β is the infection rate and a is the natural death rate of virus producing T cells. In this model ρ is the killing rate of the virus producing cells by CTL, k is the stimulation factor of CTL by infected CD4$^+$T cells, and b denotes the baseline mortality rate of CTL. We have introduced a positive feedback function of the form $\frac{ky^n}{1+ky^n}$ into our model, where n is a positive feedback control parameter.

3.3.1 Theoretical Study of the System

3.3.1.1 Equilibria

For mathematical simplicity throughout our analysis, we consider $n = 1$. Model (3.22) takes the following form:

$$\dot{x} = \lambda + rx\left(1 - \frac{x}{c}\right) - \beta xy - dx,$$
$$\dot{y} = \beta xy - ay - \rho yz,$$
$$\dot{z} = \frac{ky}{1+ky} - bz. \tag{3.23}$$

The model equation (3.23) has the following equilibria on all the coordinate planes, $E_1(x_1, 0, 0)$ and $E^*(x^*, y^*, z^*)$. The expression of the equilibrium points are

$$x_1 = \frac{c(r-d) + \sqrt{c^2(r-d)^2 + 4rc\lambda}}{2r},$$
$$x^* = \frac{a + \rho z^*}{\beta},$$
$$z^* = \frac{ky^*}{b(1+ky^*)},$$
$$g(y^*) = \pi_1 y^{*3} + \pi_2 y^{*2} + \pi_3 y^* + \pi_4 = 0, \tag{3.24}$$

where

$$\pi_1 = bk^2 c\beta^2 (\rho + ab),$$
$$\pi_2 = r\rho^2 k^2 + bkc\beta^2 \rho + bk^2\{2ar\rho - c\beta\rho(r - d)\} + 2b^2 kc\beta^2 a$$
$$\quad + b^2 k^2 \{ra^2 - \lambda\beta^2 c - \beta ca(r - d)\},$$
$$\pi_3 = bk\{2ar\rho - c\beta\rho(r - d)\} + b^2 c\beta^2 a + 2b^2 k\{ra^2 - \lambda\beta^2 c - \beta ca(r - d)\},$$
$$\pi_4 = b^2 \{ra^2 - \lambda\beta^2 c - \beta ca(r - d)\}.$$

3.3.1.2 Existence Condition

Due to complexity, we have shown the sufficient condition for the existence of a unique positive root of (3.24) numerically through Fig. 3.4. Here, we plot the cubic polynomial $g(y)$ with respect to y and for the values of the parameter set as in Table 3.2, we get three roots 21.0502, -12.9052 and -7.4548. Here $y^* = 21.0502$ is the unique positive root of (3.24).

3.3.1.3 Local Stability Analysis

Theorem 3.1 *The system (3.23) is locally asymptotically stable around E_1, if $R_0 < 1$, where $R_0 = \frac{\beta x_1}{a}$ is the basic reproductive ratio.*

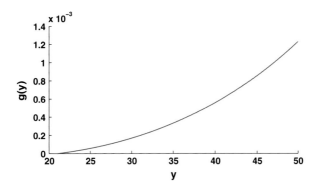

Fig. 3.4 Solution trajectory of the cubic polynomial in y with respect to $g(y)$, parameter values are as in Table 3.3

Table 3.2 Parameters used in the model (3.15)

Parameter	λ	p	T_m	d_1	β_1	d_2	β_2	β_3	s	d_3	c	d_4
Default value	10	1500	0.01	0.002	0.24	0.001	0.001	0.2	0.02	0.2	0.02	0.02

All parameter values are taken from [1–3, 8] and some are estimated

Proof The eigenvalues are given by,

$$\xi_1 = r - \frac{2rx_1}{c} - d, \quad \xi_2 = \beta x_1 - a, \quad \xi_3 = -b.$$

All the characteristic roots corresponding to E_1 will be negative if $\beta x_1 - a < 0$, i.e, $R_0 < 1$. Hence the system (3.23) is locally asymptotically stable around E_1 if $R_0 < 1$.

Theorem 3.2 *The system (3.23) is locally asymptotically stable around E^*.*

Proof The characteristic equation of system (3.23) is given by,

$$\xi^3 - (m_{11} + m_{33})\xi^2 + (m_{11}m_{33} - m_{23}m_{32} - m_{12}m_{21})\xi + (m_{12}m_{21}m_{33} + m_{11}m_{23}m_{32}) = 0,$$

$$(3.25)$$

where $m_{11} = r - \frac{2rx^*}{c} - \beta y^* - d$, $m_{12} = -\beta x^*$, $m_{21} = \beta y^*$, $m_{23} = -\rho y^*$, $m_{32} = \frac{k}{(1+ky^*)^2}$, $m_{33} = -b$. This can be expressed as

$$\xi^3 + B_1\xi^2 + B_2\xi + B_3 = 0, \qquad (3.26)$$

where $B_1 = -m_{11} - m_{33}$, $B_2 = m_{11}m_{33} - m_{23}m_{32} - m_{12}m_{21}$, $B_3 = m_{11}m_{23}m_{32} + m_{12}m_{21}m_{33}$. From the Routh–Hurwitz criterion, E^* is locally asymptotically stable if and only if

$$B_1 > 0, \ B_3 > 0, \ \text{and } B_1 B_2 - B_3 > 0,$$

where $\frac{2rx^*}{c} + \beta y^* + d - r > 0$ implies that $B_1 > 0$.

We get $\frac{\lambda}{x^*} = \frac{rx^*}{c} + \beta y^* + d - r < \frac{2rx^*}{c} + \beta y^* + d - r$. So, we can say that $\frac{2rx^*}{c} + \beta y^* + d - r > \frac{\lambda}{x^*}$, the right-hand side of this inequality is always positive, so the above condition does hold. Further, we observe that if $\frac{2rx^*}{c} + \beta y^* + d - r > 0$, then B_3 and $B_1 B_2 - B_3$ must be positive.

3.3.2 Numerical Simulation

Here, we solve the model equations (3.22) numerically in order to gain a better understanding of the previous analytical results. Initially, we choose the default values of the parameters from their reported range in various articles. The model parameters together with their default values are given in the Table 3.3. The initial values of the model variables are set to $x(0) = 150$, $y(0) = 50$ and $z(0) = 2$.

Figure 3.5 shows that the model variables $x(t)$, $y(t)$ and $z(t)$ oscillate initially before the system moves towards its stable region as time increases. It seems that as n increases from 1 to 5, the uninfected CD4$^+$T cells and CTL population density increase, whereas the infected cell population decreases.

Table 3.3 Parameters used in the model (3.23)

Parameter	λ	r	c	d	β	a	ρ	k	b
Default value	10	0.108	1500	0.01	0.002	0.24	0.2	0.001–0.005	0.02

All parameter values are taken from [1, 3, 4, 9, 10] and some are estimated

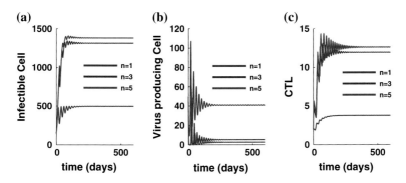

Fig. 3.5 Solution Trajectories of the system with different values of n, other parameters remain unchanged as in Table 3.3

From Figs. 3.6 and 3.7, it is clearly observed that as the drug efficacy parameter increases towards 1, for different values of n (in Fig. 3.6, $n = 2$; in Fig. 3.7, $n = 4$), the infectible cell population increases towards its maximum population density and the virus producing cells move towards extinction. However, increasing the efficacy of RTI when n is 4–8, does not produce any remarkable change in any of the cell populations being studied. Thus, selecting a highly efficacious drug during the condition when the value of n is quite high will unnecessarily increase the cost of treatment.

From our numerical study we see that the model variables $(x(t), y(t),$ and $z(t))$ oscillate initially. But the system moves towards its stable region as time increases. In a nutshell, it can be concluded that increasing the value of n always has a positive effect on uninfected cell population and keeps CTL count at a comparatively higher

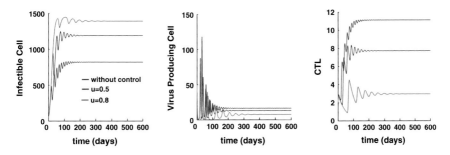

Fig. 3.6 This figure shows the system behavior for without control and with control keeping $n = 2$

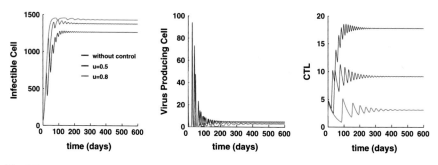

Fig. 3.7 This figure shows the system behavior for without control and with control keeping $n = 4$

magnitude whereas it exerts a negative effect on the infected cell population. This effect is much more prominent, when the values of the parameters k and ρ with n change, where the infected class count approaches zero. The destructive effect of infected cells on CTL population is evident from the decline in the CTL population as ρ increases. Another interesting outcome is that higher values of n accelerate the decline in infected cell population when ρ increases.

References

1. Bonhoeffer, S., Coffin, J.M., Nowak, M.A.: Human immunodeficiency virus drug therapy and virus load. J. Virol. **71**(3275–3278), 137 (1997)
2. Perelson, A.S., Krischner, D.E., De-Boer, R.: Dynamics of HIV infection of CD4 T cells. Math. Biosci. **114**(81–125), 118 (1993)
3. Perelson, A.S., Neuman, A.U., Markowitz, M., Leonard, J.M., Ho, D.D.: HIV 1 dynamics in vivo: viron clearance rate, infected cell life span, and viral generation time. Sci. **271**, 1582–1586 (1996)
4. Wang, L., Li, M.Y.: Mathematical analysis of the global dynamics of a model for HIV infection. Math. Biosci. **200**(44–57), 173 (2006)
5. Culshaw, R.V., Rawn, S., Spiteri, R.J.: Optimal HIV treatment by maximising immuno response. J. Math. Biol. **48**, 545–562 (2004)
6. Roy, P.K., Chatterjee A.N.: T-cell Proliferation in a mathematical model of CTL activity through HIV-1 infection. In: Proceedings of The World Congress on Engineering 2010, WCE 2010, 30 June–2 July 2010, London. Lecture Notes in Engineering and Computer Science, pp. 615–620 (2010)
7. Iwami, S., Miura, T., Nakaoka, S., Takuchi, Y.: Immune impairment in HIV infection: existence of risky and immunodeficiency thresholds. J. Theor. Biol. **260**, 490–501 (2009)
8. Wodarz, D., Nowak, M.A.: Specific therapy regimes could lead to long-term immunological control to HIV. Proc. Natl. Acad. Sci. USA **96**(25), 14464–14469 (1999)
9. Culshaw, R.V., Ruan, S.: A delay-differentianal equation model of HIV infection of CD4^{+}T-cells. Math. Biosci. **165**, 425–444 (2000)
10. Perelson, A.S., Nelson, P.W.: Mathematical analysis of HIV-1 dynamics in vivo. SIAM Rev. **41**(3–41), 122 (1999)

Chapter 4
Feedback Effect towards HIV Infection

Abstract Feedback effect has played the dominant responsibility towards HIV Infection. A negative correspondence may be expected to exist between the viral load and the rate of production of uninfected target cells. The viral load and recruitment rate of healthy CD4$^+$T cells are negatively correlated in HIV/AIDS. During low rate of infection, therapeutic interferences through using HAART, cell-specific responses of CTL are reliant on CD4$^+$T cell density. In view of above biological perspective, we have formulated mathematical models and have observed the cell dynamical behavior incorporating the negative feedback approach in detail.

Keywords Human immunodeficiency virus · Host immune system · Negative feedback control · Optimal control

4.1 Immune Cell Response to Negative Feedback Effect in HIV

A negative correlation may be assumed to exist between the viral load and the rate of production of uninfected target cells [1]. In HIV, the viral load and recruitment rate of healthy CD$^+$ T cells are negatively correlated. During low infection, therapeutic interventions using HAART, cell-specific responses of CTL is dependent on CD4$^+$ T cell density.

In view of above biological perspective with slight modification [2], the infected CD4$^+$T cells can be assumed to exert a negative feedback inhibition on the rate of formation of uninfected CD4$^+$T cells and CTL responses [3]. Thus, we have reconstructed the mathematical model [2] considering x, y, and z, which represent the uninfected CD4$^+$T cells, infected CD4$^+$T cells, and CTL response and hence, the control equations are as follows:

© Springer Science+Business Media Singapore 2015
P.K. Roy, *Mathematical Models for Therapeutic Approaches
to Control HIV Disease Transmission*, Industrial and Applied Mathematics,
DOI 10.1007/978-981-287-852-6_4

$$\dot{x} = \lambda + \frac{s_1}{k + y^m} - d_1 x - \beta xy,$$

$$\dot{y} = \beta xy - d_2 y - pyz,$$

$$\dot{z} = sy + \frac{s_2}{k + y^m} - d_3 z. \tag{4.1}$$

The system will be analyzed with the following initial conditions : $x(0) > 0$, $y(0) > 0$, $z(0) > 0$ and we denote

$$R_+^3 = \{(x, y, z) \in R^3, x \geq 0, y \geq 0, z \geq 0\}.$$

We have assumed s as the rate of stimulation of CTL. Here, s_1 and s_2 are growth terms and m is defined as a feedback factor. We also assume k as the host-virus interaction coefficient.

4.1.1 Theoretical Analysis

4.1.1.1 Existence Condition

In the present situation (4.1), three types of equilibria can exist:

(i) disease-free equilibrium $E_1(\frac{\lambda k + s_1}{k d_1}, 0, 0)$,

(ii) equilibrium $E_2(\frac{\lambda k + s_1}{k d_1}, 0, \frac{s_2}{k d_3})$, and

(iii) endemic equilibrium $E^*(x^*, y^*, z^*)$ where,

$$x^* = \frac{1}{\beta}[d_2 + \frac{p}{d_3}(sy^* + \frac{s_2}{k + y^{*m}})], \quad z^* = \frac{1}{d_3}(sy^* + \frac{s_2}{k + y^{*m}})$$

and

$$ps\beta y^{*m+2} + (psd_1 + d_2 d_3 \beta)y^{*m+1} + d_3(d_1 d_2 - \lambda \beta)y^{*m} + psk\beta y^{*2}$$
$$+ \{pskd_1 + \beta(d_2 d_3 k + ps_2)\}y^* - \{\beta(\lambda d_3 k + s_1 d_3) - d_1(d_2 d_3 k + ps_2)\} = 0.$$

If the above last equation has only one positive root, then the steady state exists and uniqueness of the steady state is confirmed by Descartes rule of sign. Thus, we get the condition stated below, for which E^* always exists.

$$d_1 d_2 > \lambda \beta, \quad d_3 \beta(\lambda k + s_1) > d_1(ps_2 + d_2 d_3 k). \tag{4.2}$$

4.1.1.2 Boundedness of the System

To verify the boundedness of the system, the following lemma is very much useful.

Lemma 4.1.1 *For $x(t)$ satisfying $x'(t) < c - q(\phi)x(t)$, where c is a constant and $q(\phi)$ is independent of x and t, $x(0) < \frac{c}{q(\phi)} \Rightarrow x(t) < \frac{c}{q(\phi)}$, for all t.*

Proof See Lemma 4.1.1 of [4] for the proof.

For the system (4.1),

$$x(t) + y(t) \leq \frac{\lambda k + s_1}{kd_1}, \tag{4.3}$$

$\Rightarrow x(t) \leq \frac{\lambda k + s_1}{kd_1}$ and $y(t) \leq \frac{\lambda k + s_1}{kd_1}$. Now,

$$\dot{z}(t) \leq s\left(\frac{\lambda k + s_1}{kd_1}\right) + \frac{s_2}{k} - d_3 z \Rightarrow z(t) \leq \frac{s(\lambda k + s_1) + s_2 d_1}{kd_1 d_3}. \tag{4.4}$$

Hence, the system is bounded in the region Ω defined below:

$$\Omega = \{(x(t), y(t), z(t)) \in R_+^3, \ x(t) \leq \frac{\lambda k + s_1}{kd_1}, \ y(t) \leq \frac{\lambda k + s_1}{kd_1},$$
$$z(t) \leq \frac{s(\lambda k + s_1) + s_2 d_1}{kd_1 d_3}\}.$$

4.1.2 Stability of the System

The eigen values are $-d_1$, $\frac{\beta(\lambda k + s_1)}{d_1 k} - d_2$ and $-d_3$, respectively.
Thus, we get the basic reproductive ratio $R_0 = \frac{\beta(\lambda k + s_1)}{d_1 d_2 k}$. It follows that E_1 always exists and locally stable if $R_0 < 1$ and unstable if $R_0 > 1$.

Proposition 4.1.2 *The disease-free equilibrium E_1 corresponds to the maximum level of healthy uninfected CD4$^+$T cells in the absence of virus when the population of infected cells or CTL is nil. Here E_1 is attained, when $R_0 < 1$. Furthermore, if $R_0 < 1$, then E_1 is local asymptotically stable. If $R_0 > 1$, then the system E_1 becomes unstable.*

The eigen values are $-d_1$, $\frac{\beta(\lambda k + s_1)}{d_1 k} - d_2 - \frac{ps_2}{kd_3}$ and $-d_3$ respectively. Now all the eigenvalues have negative real parts if $R_0 < 1 + \frac{ps_2}{kd_3}$. Hence, we establish the proposition.

Proposition 4.1.3 *Equilibrium state E_2 exists, when R_0 varies between 1 and $1 + \frac{ps_2}{kd_2d_3}$, i.e., if $1 < R_0 < 1 + \frac{ps_2}{kd_2d_3} = 1 + R_1$, then E_2 is local asymptotically stable. If $R_0 > 1 + \frac{ps_2}{kd_2d_3}$, then the system E_2 is unstable. In other words, a condition for existence of E_2 holds, when immune response is highly active in killing the infected cells, i.e., the host system has a tendency to eliminate infection on its own and is yet to attain endemic equilibrium.*

The characteristic equation for $J(E^*)$ is $\rho^3 + a_1\rho^2 + a_2\rho + a_3 = 0$, where

$$a_1 = d_1 + d_3 + \beta y^* > 0,$$
$$a_2 = d_3(d_1 + \beta y^*) + spy^* + \beta^2 x^* y^* + \frac{my^{*m}}{(k+y^{*m})^2}(\beta s_1 - ps_2) \text{ and}$$
$$a_3 = d_3\beta^2 x^* y^* + (d_1 + \beta y^*)spy^* + \frac{my^{*m}}{(k+y^{*m})^2}\{d_3\beta s_1 - (d_1 + \beta y^*)ps_2\} > 0.$$

From Routh–Hurwitz condition, the necessary and sufficient condition for local asymptotical stability of the steady state is $a_1a_2 - a_3 > 0$.

Proposition 4.1.4 *The system E^* is stable if*

(i) $1 + \frac{ps_1}{d_2d_3k} < R_0 < 1 + \frac{s_1}{\lambda k} = 1 + R_2$ and
(ii) $a_1a_2 - a_3 > 0$ are satisfied.

Thus, we can conclude that the endemic equilibrium E^* exists, when all the three components considered in host-virus dynamics possess positive values, and the system attains the infected equilibrium state. Moreover, it has been observed that E^* does not exist for very large values of k or host-virus interaction coefficient. High viral load affects the source of T cells, kills uninfected and infected $CD4^+$ T cells and results in the immunological gap due to deficiency of CTL stimulation (Fig. 4.1).

4.1.3 Numerical Analysis

We now numerically illustrate the change of the stability due to the negative feedback factor. We have chosen the initial condition for the parameters given as in

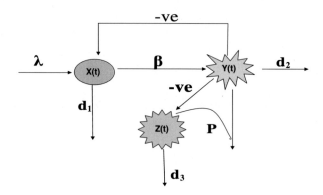

Fig. 4.1 Schematic diagram of mathematical model (4.1)

Table 4.1 Parameters used in the model (4.1)

Parameter	λ	d_1	β	d_2	p	s	d_3
Default value	$1-10$	$0.007-0.1$	$0.00025-0.5$	$0.2-0.3$	0.002	$0.1-1$	$0.1-0.15$

All parameter values are taken from [2, 5]

Table 4.1. The initial values of the model variables are considered as $x(0) = 1000$, $y(0) = 100$, $z(0) = 10$ [6], and the cell population is expressed as per mm^3. It should be noted that the asymptotic time series solutions of the model equation do not depend on the choice of the initial values of the model variables. Variation of the parameter p is restricted by the condition $\frac{ps}{d_3} \sim 0.01 - 0.05$ [2] which implies that CTL activity is limited up to a certain value because of the negative feedback effect exerted by high viral load on the stimulation and persistence of HIV-specific immune response after which immune system can no longer trigger off to fight against foreign antigens. The parameters s and d_3 are as mentioned in the Table 4.1. Figure 4.2 shows the existence and stability conditions for the systems E_1, E_2, and E^*. Here, we plot the basic reproductive ratio R_0 with respect to k. The zone above the red line indicates the zone, where there is no possibility of existence of equilibrium of any type because of total breakdown of the host immunity system due to large value of k. Thus the infection persists throughout without any chance of remission. Figure 4.3 reflects that numerical value of eradication threshold remains above one. It possesses similar magnitude if rate of production of CD4$^+$T cells is increased keeping host-virus interaction coefficient constant at low values and rate of infection considerably high. Thus, the disease outbreaks because of the availability of target uninfected cells susceptible to infection, high infectivity of virus which is less than the required CTL population for containment of infection. Eradication is possible ($T_0 < 1$) for low values of β even if both k and λ are increased because CTL response is effective at this stage. However, with the continuous increase of β, a point will come when eradication is no longer possible because of collapse of immune system. Thus, it can be concluded that k acts as a determinant in deciding the conditions for existence of eradication threshold or disease outbreak and CTL stimulation depends on β.

Fig. 4.2 Phase diagram showing the basic reproduction ratio for the system (4.1) as a function of k

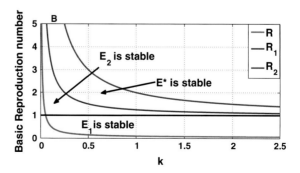

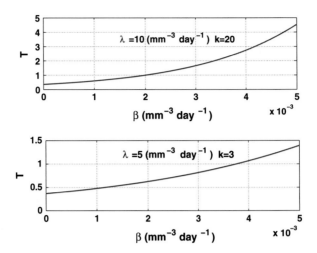

Fig. 4.3 Phase diagram showing the outbreak condition of the system (4.1) as a function of β

Fig. 4.4 Schematic
explanation of mathematical
model (4.1)

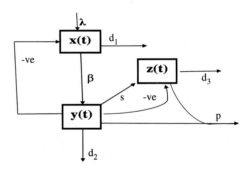

Therapeutic intervention, which will interfere with interaction between host cells
and virus can theoretically lead to disease-free condition when eradication threshold
T is less than one (Fig. 4.4).

4.1.4 Discussion

In the present study, the concept of the host-virus interaction coefficient k has been
introduced in delineating the complexities of host-virus interaction dynamics. For
proper characterization of long-term dynamics, it becomes essential to incorporate
negative feedback control effect induced by high viral load on stimulation of HIV-
specific CTL-mediated immune response. During primary stage of infection, when
the viral load is substantially low and CTL response is yet to develop in a full-fledged
manner, negative feedback effect is negligible. The existence conditions of stability

of three equilibria are focused in this section depending on different values of the basic reproductive ratio R_0. Effect of negative feedback control by high viral load on CTL stimulation is observed in the relationship between β and T when threshold conditions for eradication (or infection persistence) is established.

4.2 Negative Feedback Effect in HIV Progression

As uninfected CD4$^+$T cells decreases, host-virus interaction decreases and CTL response fails to develop, resulting in disease progression. When the rate of infection is high and m attains a much higher value. It exerts a significant effect on host-virus interaction and k^m becomes negligibly small. However, this situation happens only if the viral load is sufficiently high, when the infected cell population is considerably greater. Consequently, the final effect is lowering in the number of uninfected cell count. The above statements justify the introduction of k^m in the model in lieu of k (4.1), for proper characterization of long-term HIV dynamics with respect to existence of negative feedback effect in the system. Thus, we reconstruct the mathematical model (4.1) considering x, y, and z which represent the uninfected CD4$^+$T cells, infected CD4$^+$T cells and CTL response and hence the control equations are as follows:

$$\dot{x} = \lambda + \frac{s_1}{k^m + y^m} - d_1 x - \beta x y,$$
$$\dot{y} = \beta x y - d_2 y - p y z,$$
$$\dot{z} = s y + \frac{s_2}{k^m + y^m} - d_3 z. \tag{4.5}$$

The system needs to analyze with the following initial conditions: $x(0) > 0$, $y(0) > 0$, $z(0) > 0$ and we denote

$$R_+^3 = \{(x, y, z) \in R^3, x \geq 0, y \geq 0, z \geq 0\}.$$

Here, we assume s as the rate of stimulation of CTL responses. Here, s_1 and s_2 are growth terms and m is defined as feedback factor. We also assume k as host-virus interaction coefficient. Considering the non-dimensional quantities

$$X = xk^{-1},\ Y = yk^{-1},\ Z = zk^{-1},\ T = tk\beta,\ l = \frac{\lambda}{\beta k^2},\ A = \frac{s_1}{\beta k^{m+2}},\ B = \frac{s_2}{\beta k^{m+2}},$$
$$\delta_1 = \frac{d_1}{\beta k},\ \delta_2 = \frac{d_2}{\beta k},\ \delta_3 = \frac{d_3}{\beta k},\ \alpha_1 = \frac{p}{\beta},\ \alpha_2 = \frac{s}{\beta k}.$$

The dimensionless form of the model becomes,

$$\dot{X} = l + \frac{A}{1 + Y^m} - \delta_1 X - XY,$$
$$\dot{Y} = XY - \delta_2 Y - \alpha_1 YZ,$$
$$\dot{Z} = \alpha_2 Y + \frac{B}{1 + Y^m} - \delta_3 Z. \tag{4.6}$$

4.2.1 Steady-State Analysis

The system (4.6) has the following steady state. The first one is disease-free equilibrium given by $E'(\frac{l}{\delta_1}, 0, 0)$ and another is infected equilibrium $E^*(X^*, Y^*, Z^*)$. The
infected steady state E^* of the system (4.6) is obtained from

$$X^* = \delta_2 + \frac{\alpha_1}{\delta_3}(\alpha_2 Y^* + \frac{B}{1 + Y^{*m}}), \quad Z^* = \frac{1}{\delta_3}(\alpha_2 Y^* + \frac{B}{1 + Y^{*m}})$$

and

$$\alpha_1 \alpha_2 Y^{*m+2} + (\alpha_1 \alpha_2 \delta_1 + \delta_2 \delta_3) Y^{*m+1} + (\delta_1 \delta_2 \delta_3 - l\delta_3) Y^{*m} + \alpha_1 \alpha_2 Y^{*2}$$
$$+ (\alpha_1 \alpha_2 \delta_1 + \delta_2 \delta_3 + \alpha_1 B) Y^* + (\alpha_1 \delta_1 B - l\delta_3 - \delta_3 A) = 0. \tag{4.7}$$

If the last Eq. (4.7) has only one positive root, then the steady state exists and uniqueness of the steady state is confirmed by Descartes rule of sign. Thus we get the
condition (4.8) stated below, for which E^* always exists.

$$\frac{pd_1 s_2}{d_3(\lambda k^m + s_1)} < \beta < \frac{d_1 d_2}{\lambda}. \tag{4.8}$$

Remark: From the above existence condition we can conclude that if the rate at which
the uninfected T cells become infected by the virus particle is restricted then only
infected steady state exists. We can also state that the uninfected T cell population
and CTL response are both affected by the feedback factor if and only if m increases
reflected by decline in uninfected T cells and CTL response.

4.2.1.1 Stability of the System

Here, we consider the basic reproductive ratio R_0, which means the average number
of secondary infection caused by a single infected T cell in an entirely susceptible T

cell. Here $R_0 = \frac{\lambda \beta}{d_1 d_2}$. For E', the Jacobian matrix becomes J' where the eigen values for J' are $-\delta_1$, $-(\frac{l}{\delta_1} + \delta_2)$ and $-\delta_3$.

Thus we get the proposition.

Proposition 4.2.1 *If $R_0 < 1$, then E' is local asymptotically stable. If $R_0 > 1$, then the system E' is unstable.*

The Jacobian matrix for E^* is J^*. The characteristic equation for $J(E^*)$ is $\rho^3 + a_1 \rho^2 + a_2 \rho + a_3 = 0$, where

$$a_1 = \delta_1 + \delta_3 + Y^* > 0,$$
$$a_2 = \delta_1 \delta_3 + \delta_3 Y^* + X^* Y^* + \alpha_1 \alpha_2 Y^* + \delta^* (A Y^* - B \alpha_1 Y^*) > 0,$$
$$a_3 = Y^* [\{\delta_3 X^* + \alpha_1 \alpha_2 (\delta_1 + Y^*)\} + \delta^* \{A \delta_3 - \alpha_1 B (\delta_1 + Y^*)\}] > 0$$

and $\delta^* = \frac{m Y^{*m-1}}{(1+Y^{*m})^2} > 0$.

From Routh–Hurwitz criterion, the necessary and sufficient condition for locally asymptotically stability of the steady state is $a_1 a_2 - a_3 > 0$

$$\Rightarrow \delta_1 \delta_3 (\delta_2 + \delta_3 + 2Y^*) + \delta_3 Y^* (\delta_3 + \alpha_1 \alpha_2 + Y^*) + X^* Y^* (Y^* + \delta_1) + \delta^* Y^* (A \delta_1$$
$$+ A Y^* - B \alpha_1 \delta_3) > 0.$$

Proposition 4.2.2 *The system E^* is stable if (i) $R_0 > 1$ and (ii) $a_1 a_2 - a_3 > 0$ are satisfied.*

4.2.2 Numerical Analysis

We now numerically illustrate the change of the stability due to varying the time delay. We choose the initial conditions of the parameters given as in Table 4.2. At $t = 0$ the values of the model variables are considered as $x(0) = 1000$, $y(0) = 100$, $z(0) = 10$. It should be noted that the asymptotic time series solutions of the model equation do not depend on the choice of the initial values of the model variables. Variation of the parameter p is restricted by the condition $\frac{ps}{d_3} \sim 0.01 - 0.05$ [2]. The parameter s and d_3 are as mentioned in the Table 4.1.

Figure 4.5 (left panel) shows the existence and stability conditions for the system E'. Here, we plot the basic reproductive ratio R_0 with respect to β and d_2. It is easily seen that for reducing value of β, basic reproductive ratio R_0 reduces proportionately.

Table 4.2 Parameters used in the model (4.5)

Parameter	λ	d_1	β	d_2	p	s	d_3
Default value	$1 - 10$	$0.007 - 0.1$	$0.00025 - 0.5$	$0.2 - 0.3$	0.002	$0.1 - 1$	$0.1 - 0.15$

All parameter values are taken from [2, 7–9]

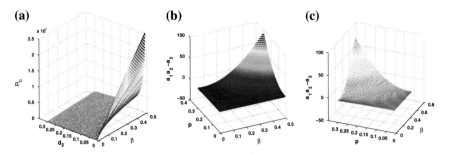

Fig. 4.5 *Left panel* **a** Phase plane for the condition of existence of the stability of the uninfected steady state E'. *Center panel* **b** Mesh diagram showing the existence condition of E^* for β, p, and $m = 1$. *Right panel* **c** Mesh diagram showing the existence condition of E^* for β, p and $m = 2$.

If we restrict $\beta < 0.1$, then it shows that $R_0 < 1$, which reflects the stability of the system. But if d_2 increases, then $R_0 < 1$ regardless of large value of β.

From Fig. 4.6 (right panel) we observe that if $m \geq 4$, the stationary point becomes unstable which implies that infected steady state is disturbed and the viral count starts increasing. It contradicts in vivo viral replication because high viral load is known to exert a negative feedback effect on the supply of fresh target cells and attains a quasi-steady state of viral population. In Fig. 4.6a, b we see that for $m = 1$, $a_1 > 0$, when $\beta < 0.004$. But if $\beta > 0.004$, $a_1 < 0$. Thus, if the rate of infection is restricted, then E^* remains stable. But when $m = 2$ (Fig. 4.6c, d) the system E^* remains stable for $\beta > 0.004$.

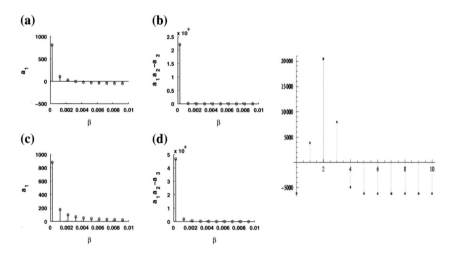

Fig. 4.6 *Left panel* Phase plane for the condition of existence of the stability of the infected steady state E^*. *Right panel* Figure shows that when $m \geq 4$ the system becomes unstable. When $m < 4$ the system remains stable. That means for large compatibility the system becomes unstable

If β lies above a threshold value (i.e., 0.004) and feedback factor is of lower magnitude ($m = 1$), it becomes practically impossible to exert immunological restriction on the disease progression. It is occurred because of rapid infection of available few uninfected cells, failure of establishment of effective, sustained CTL response in presence of viral antigen, and a substantial decline in the count of uninfected T cells in comparison to the number of infected cells. The system is yet to attain persistent infection equilibrium.

However, if β lies below the threshold value, it will take longer time to infect uninfected cells. Immune system is strong enough to control the infection because of the development of effective CTL response. Viral load and the number of infected cells are on a decline leading to the generation of disease-free equilibrium. Infected steady state can be attained if the feedback factor increases ($m = 2$), when $\beta > 0.004$. Now the immune system is totally impaired due to weakening of HIV-specific CTL-mediated responses and further viral replication is inhibited due to depletion of uninfected CD4$^+$T cells and death and destruction of T cell progenitors.

References

1. Huang, Y., Lu, T.: Modeling long-term longitudinal HIV dynamics with application to an AIDS clinical study. Ann. Appl. Stat. **2**, 1384–1408 (2008)
2. Bonhoeffer, S., Coffin, J.M., Nowak, M.A.: Human immunodeficiency virus drug therapy and virus load. J. Virol. **71**(3275–3278), 137 (1997)
3. McCune, J.M.: The dynamics of CD4$^+$T-cell depletion in HIV diseases. Nature **410**, 974–979 (2001)
4. Smith, R.J., Wahl, L.M.: Distinct effects of protease and reverse transcriptase inhibition in an immunological model of HIV-1 infection with impulsive drug effects. Bull. Math. Biol. **66**(5), 1259–1283 (2004)
5. Perelson, A.S., Krischner, D.E., De-Boer, R.: Dynamics of HIV infection of CD4 T cells. Math. Biosc. **114**(81–125), 118 (1993)
6. Culshaw, R.V., Rawn, S., Spiteri, R.J.: Optimal HIV treatment by maximising immuno response. J. Math. Biol **48**, 545–562 (2004)
7. Perelson, A.S., Neuman, A.U., Markowitz, J., Leonard, M., Ho, D.D.: HIV 1 dynamics in vivo: viron clearance rate, infected cell life span, and viral generation time. Science **271**, 1582–1586 (1996)
8. Perelson, A.S., Nelson, P.W.: Mathematical analysis of HIV-1 dynamics in vivo. SIAM Rev. **41**(3–41), 122 (1999)
9. Roy, P.K., Chatterjee A.N.: T-cell proliferation in a mathematical model of CTL activity through HIV-1 Infection. In: Lecture Notes in Engineering and Computer Science: Proceedings of The World Congress on Engineering 2010, WCE 2010, pp. 615–620. London, U.K. 30 June–2 July 2010

Part II
Control-Based Therapeutic Approach

In this part we discuss the effect of delay for controlling the disease during different stages of infection for different dynamics. Also, it has been observed that there is delay in the infection process because there is a required time needed for a newly infected cell to start producing HIV virus [1]. During this infection process there is an intercellular "eclipse phase" or "latency phase" during which cell is infected but does not begin to produce new virus. This idea might be modeled by either delay or by an explicit latency class [1].

Culshaw and Ruan [2] studied the basic model of HIV infection considering the logistic growth and an intercellular delay. They did not find any significant condition that led to hopf bifurcation. Herz et al. [3] assumed that cells become productively infected after τ units of initial infection. They observe that on considering intercellular delay viral clearance rate was estimated but did not change the amount of productively infected cell. Culshaw et al. [4] considered cell to cell spread of HIV tissue culture (in vitro) and they modeled the intercellular eclipse phase by a gamma distribution. Perelson et al. assumed that there are two types of delay which are of pharmacological origin that occurs between the administration of drug and the intercellular delay [5–8]. Several mathematical models have been studied and the intracellular delay has marginal effects that are analytically justifiable. It has been observed that the immune system of the infected patients becomes highly risky for developing opportunistic disease when the healthy $CD4^+T$ cell count falls below a critical value. Thus, during the treatment of HIV patients the virus production rate and the clearance rate of free virus escalate. But in reality, controlling the model parameters is quite difficult, while controlling the drug dosage is more realistic. Thus, our main focusing areas are the effect of different drug therapy using control theoretic approach.

HIV grows weaker in the immune system due to its infection, but the virus still works inside the immune system. Here, the role of the Antigen-Presenting Cell (APC) of an HIV infected individual is thus important as it indicates precursor Cytotoxic T Lymphocytes to differentiate into killer T-cells known as effectors CTL

[9]. On the other hand, killer T-cells instigate to destroy infected $CD4^+T$ cells from where new infectious viruses are born. This implies that there have been several complications in the immune system for using drugs or, more precisely, it can be said that the infection process is quite a complicated interaction process between different cells, virus, and drugs [10].

Mathematical models of drug treatment dynamics suggest that the CTL response could be maintained or even increased by combining drug therapy with vaccination [11–12]. When drug is administered in an HIV infected individual, CTL is stimulated and it acts against the infected $CD4^+T$ cells. In recent years, the antiretroviral therapy for HIV positive patients has largely improved. Different drug therapy is administered for different stages of the HIV infected patients. There are more than twenty Food and Drug Administration (FDA) recommended anti-HIV drugs available. Since the primary receptor for HIV is CD4 receptor on T cell and the hallmark feature of AIDS is total impairment of host immune system with depletion of $CD4^+T$ cell pool, the ultimate objective of any successful therapeutic intervention will be replenishment of $CD4^+T$ count, reconstitution of immune system, and eradication of virus from the system. Administration of Highly Active Anti Retroviral Therapy (HAART), comprising of a combination of Reverse Transcriptase Inhibitors (RTIs) and Protease Inhibitors (PIs), has substantially improved the well-being and life-expectancy of HIV-positive patients but has failed in certain aspects that can be overcome by adoption of complementary strategies. Most of these therapeutic agents falls in two categories: reverse transcriptase inhibitors (RTI: AZT, ddI, ddC, D4T, 3TC, Delavirdine, Nevirapine, Bacavir, Succinate, and Efavireenz) and protease inhibitors (PI: Ritonavir, Saquini, Indinavir, and Nelfinavir) [13]. The reverse transcriptase inhibitors are used to prevent HIV RNA from being converted into DNA, thus blocking integration of the viral code into the target cells [13, 14]. Reverse Transcriptase Inhibitors (RTIs) prevent HIV from infecting cells or put a stop to infection of new cells. The protease inhibitors efficiently reduce the number of infectious virus particles released by an infected cell [13, 14]. Protease Inhibitors (PIs) prevent the production of new infectious virus by infected cells [11, 12], thus causing the virus to be unable to infect helper-T cells. Since, the Highly Active Anti Retroviral Therapy (HAART) is the most effective treatment nowadays, which is mainly a combination of three or more different drugs of RTI and PI. This type of treatment is effective for preventing new infection and killing or halting the virus. HAART can achieve only partial immune reconstitution and viral reservoir of latently infected cells which remain almost completely unaffected. Moreover, in long-term therapy, these anti-HIV agents become ineffective due to mutation in virus.

It has also been observed that a long term use of these highly expensive drug therapies produces results with many complications. After taking these drugs, patients may suffer from harmful pharmaceutical side effects such as cardiovascular, lactic acidosis, etc. [15]. As a result, the patient may die of other diseases. This type of treatment for HIV does not improve the immune system to help the body to fight against any other infection. Nowadays many clinical laboratories keep a systematic data records of patient treatment courses with respect to effectiveness and results. But these records provide an incompatible indication as to which is better: early treatment (defined as $CD4^T$ cell counts between $500 \approx 200$ mm^{-3} of blood) or treatment at a later stage (below 200 mm^{-3}). "Better" here is based on overall health of patient (i.e., side effects) and a preservation or amplification in the $CD4^+T$ cells count [16]. To activate the immune system, cytokines play an important role. Cytokines are protein hormones and Interleukin-2 (IL-2) is the main cytokine for which the immune system is activated [17]. IL-2 is mainly produced from $CD4^+T$ cells and $CD8^+T$ cells and it acts on the same cell from which it produces. IL-2 also activates the $CD4^+T$ cells and partially $CD8^+T$ cells. In spite of $CD4^+T$ cell and $CD8^+T$ cell differentiation, IL-2 also activates the latently infected cells. Since, HIV infected patients immune system becomes weak, IL-2 does not act accurately as its production is impaired [17]. Antiretroviral therapy can partially stimulate IL-2 production, but it does not work accurately for a long period of time. Thus, immunotherapy with IL-2 is a more effective treatment to fight against HIV infection.

The control theoretic concepts have been considered important in a wide variety of disciplines. Since, too large dosage may not be desirable for patients while too small dosage may be ineffective as therapy for the recommended therapeutic agents. Also in a number of situations, an initial dose of drug is given to be followed with an intermittent interval regularizing the maintained dosage. Optimal treatment strategies can decrease the possibility of virus mutation, pharmaceutical side effects, and expensive medication burden. To avoid complication due to toxic effects of the drug, adequate amounts of drug in a body compartment should be maintained. To avoid the hazard of side effect of drug dose, our main aim is to find out the optimal drug dosage. Here the drug input is the control and it is through the knowledge of their size that one has a partial way of influencing the drug response behavior among patients [18]. The basic equation of the optimal control theory may be derived by a different approach, which comprises Pontryagin's Minimum Principle [19].

The concepts of optimal control theory have been important in an extensive variety of disciplines. Over the years, the theory has been developed and extended. Simultaneously, the application and necessity of the theorem is much important for the drug treatment in different treatment managements. Thus we have used the theory and with the help of this theorem we have derived the optimal drug dosage through which better treatment can be achieved. In late 1950, Lev Semenovic Pontryagin and his co-workers developed the Optimal control theory and generalized the theory using the calculus of variations [19]. Pontryagin developed the maximum principle for optimal control of finite dimensional problems governed by ordinary differential equations (ODEs). Here variables are divided into two classes, state variables and control variables. The state variables are governed by first-order differential equations. The trajectories of the state variables are inclined directly by the control variables. It is to be mentioned that the number of control variables may not be equal to the number of state variables. The control variables govern the state variables, which are chosen to maximize an objective functional, respecting the desired goal.

In the control therapeutic approach a large number of mathematical models have been proposed by eminent researchers [20–22, 15]. Kirschner et al. [16] considered the interaction of HIV with immune systems and included a chemotherapy control to determine the optimal control dynamics of the drug dosage. Fister worked out on the same model [21] and examined the optimal control which represented a percentage effect of chemotherapy on the interaction of $CD4^+T$ cells with HIV and found the minimized systemic cost based on the drug dosage. The effect of combination drug therapy of RTI and IL-2 on the interaction of HIV and $CD4^+T$ cells have been studied by Joshi [23]. We have come to know about the structured treatment interruption (STI) from the paper of Admas et al. [13], where he showed that treatment reduced the pharmaceutical side effects in HIV treatment. Culshaw et al. [24] established an optimal control model of HIV treatment using a single drug. They show that immune response may be rejuvenated by optimal treatment strategies.

Inspite of the success of HAART, long-term control in disease progression is problematic and the immune system remains impaired [25]. Also, the latently-infected $CD4^+T$ cells and other cells carry the replication process to form free HIV. HIV-specific $CD8^+T$ lymphocytes play a vital role to control HIV replication in vivo. These $CD8^+T$ cells are the precursors for anti-HIV cytotoxic T lymphocytes (CTLs) that destroy HIV-producing cells. Depletion of $CD8^+T$ cells causes a significant increase in HIV replication [26]. Intermittent administration of HAART and immune activators such as IL-2 has been proposed as a possible strategy for control of viral replication [27]. Kirschner, Webb formulate the mathematical model of IL-2 treatment of HIV infection [17] and they found that this type of therapy can be successful in delaying AIDS progression. They also found that immunotherapy is most beneficial for raising $CD4^+T$ cell count. Gumel also established a mathematical model including IL-2 and HAART [28], and showed that the development of a potent vaccine stimulates the proliferation of HIV-specific $CD4^+T$ cells and CTLs. They also discussed and developed a new class of anti-HIV drugs that target and eliminate non-$CD4^+T$ cell HIV reservoirs. Roy and Chatterjee [29] have shown that when the immune responses are high, less medication is needed to control and regulate infection. Our mathematical models, also reflect that optimal treatment reduces the period of time while the immune response of the uninfected $CD4^+T$ cells takes over.

The possibility of immunotherapy to correct individual HIV-driven immune alteration by exploiting the specific effects of different immunomodulants like IL-2 on T cell dynamics is indeed a fascinating and novel perspective in HIV treatment. Increasing evidence favors co-administration of HAART and IL-2 following an optimal treatment schedule leading to selective expansion of immune system and near extinction of viral population from the system [30, 31]. Many models assume that the drug was widely available within the body and the average efficacy varied between 0 (complete drug failure) and 1 (complete inhibition of the virus). These models have the advantage of making models simpler, allowing for greater generality. However, the disadvantage is that the dynamics of drug behavior are ignored. But the control of cellular infection along with activation of immune system, introducing of optimal theory, or drug perfect adherence is yet to be explored.

The theory of impulsive differential equation is much richer than the corresponding theory of differential equation without impulse effect. The theory is interesting in itself and it is easy to see that it will assume greater importance in the near future since the application of the theory in various fields is also increasing. Thus impulsive differential equation, involving impulsive effects, appears as a natural description of observed evolution phenomena of several real-valued problems.

The theory of impulsive differential equations is relatively new. The main results are found in Lakshmikantham [32], Bainov and Simeonov [33]. The theory has found applications in many areas where evolutionary processes undergo rapid changes at certain times of their development. In the mathematical simulation of such processes, the duration of this rapid change is ignored and instead it is assumed that the process changes its state instantaneously. The theory highlighted many applications in different physical systems like mechanics, radio engineering, control theory, and biotechnology. As a case, it is applied to model self-cycling fermentation process by a series of impulsive differential equations, which describes the inherent properties with robust resolute. Most of the theory of impulsive differential equations has been carried out for non-autonomous equations, where the times of the impulsive effect are fixed. This makes the analysis relatively straightforward, although it should be noted that the detail is still quite complicated. The equations modeling self-cycling fermentation are autonomous, with variable (and non-explicit) moments of impulse.

The study of drug dynamics is determined by impulsive differential equation. Perfect or imperfect drug adherence to HIV infection can facilitate the development of resistance. Thus in recent years the effects of perfect adherence to antiretroviral therapy have been studied by impulsive differential equation [34–39]. Using this method, the dosing period and threshold values of dosage can be obtained more precisely. Also, the effect of maximal acceptable drug holidays can be found by the study of impulsive differential equations [40]. Impulsive differential equations result if drug effect as well as the metabolites are assumed to decay with time in an exponential manner during each cycle and is assumed to change instantaneously. The dosing parameter for different drug doses can result in either implicit or explicit models [35–38].

The drug administered is first dissolved into the gastrointestinal tract. The drug is then absorbed into so-called apparent volume distribution and finally eliminated from the system. Pharmacokinetics is a discipline that describes and predicts the time-course of drug concentrations in body fluids. Compartment models are widely used to describe the drug absorption, distribution, and elimination in human and animal body [41, 42]. Body is described through one or multiple compartments, and from which drugs could be absorbed, transferred, and eliminated according to the simple kinetic rate expressions. Here we also deal with the drug dynamics to control the disease.

References

1. Nelson, Murray, J.D., Perelson, A.S.: Oscillatory viral dynamics. Math Bio. Sc. **63**. 111 (2000)
2. Culshaw, R.V., Ruan, S.: A delay -differential equation model of HIV infection of $CD4^+T$-cells. Math. Biosci. **165**, 425–444 (2000)
3. Herz, A.V.M., Bonhoeffer, S., Anderson, R.M., May, R.M., Nowak, M.A.: Viral dynamics in vivo; limitations on estimates of intracellular delay and virus decay. Proc. Natl. Acad. Sci. USA. **93**, 7247–7251 (1996)
4. Culshaw, R.V., Ruan, S., Webb, G.: A mathematical model of cell to cell spread of HIV-1 that includes a time delay. J. Math. Biol. **46**, 425–444 (2003)
5. Perelson, A.S.: Mathematical and Statistical Approaches to AIDS Epidemiology, p. 116. Springer, Berlin (1989)
6. Perelson, A.S., Krischner, D.E., De-Boer, R.: Dynamics of HIV infection of $CD4^+T$cells. Math. Biosc. **114**, 81–125.118 (1993)
7. Perelson, A.S., Neuman, A.U., Markowitz, M., Leonard, J.M., Ho, D.D.: HIV 1 dynamics in vivo: viron clearance rate, infected cell life span, and viral generation time. Science. **271**, 1582–1586.120 (1996)
8. Perelson, A.S., Nelson, P.W.: Mathematical Analysis of HIV-1 Dynamics in Vivo. SIAM Review. **41**, 3–41 (1999)
9. Altes, H.K., Wodarz, D., Jansen, V.A.A.: The dual role of $CD4^+T$ Helper Cells in the Infection Dynamics of HIV and Their Importance for Vaccination. J. Theor. Biol. **214**, 633–644 (2002)
10. Skim, H., Han, S. Ju., Chung, C.C., Nan S.W., Seo, J.H.: Optimal Scheduling of Drug Treatment for HIV Infection. Int. J. Control Autom. Syst. **1**(3), 282–288.143 (2003)
11. Wodarz, D., Nowak, M.A.: Specific therapy regimes could lead to long-term immunological control to HIV. Proc. Natl. Acad. Sci. USA. **96**(25), 14464–14469.169 (1999)
12. Wodarz, D., May, R.M., Nowak, M.A.: The role of antigen-independent persistence of memory cytotoxic T lymphocytes. Int. Immunol. V**12**(A), 467–477.171 (2000)
13. Cao, Y., Qin, L., Zhang, L., Safrit, J., Ho, D.D.: Virologic and immunologic characterization of long term survivors of human immunodeficiency virus type. N. Engl. J. Med. **26**, 1–8 (1995)
14. Kim, W.H., Chung, H.B., Chung, C.C.: Optimal Switching in Structured Treatment Interruption for HIV Therapy. Asian J. Control. **8**(3), 290–296 (2006)
15. Kwon, H.D.: Optimal treatment strategies derived from a HIV model with drug-resistant mutants. Application in Mathematics and Computation. **188**, 1193–1204 (2007)
16. Kirschner, D., Lenhart, S., Serbin, S.: Optimal Control of the Chemotherapy of HIV. J. Math. Biol. **35**, 775–792 (1997)
17. Kirschner, D.E., Webb, G.F.: Immunotherapy of HIV-1 infection. J. Biol. System 6(1), 71–83. population dynamics. Math. Biosci. **170**, 187–198 (1998)
18. Swan, G.M. Application of Optimal Control Theory in Biomedicine. 135 (1984)
19. Pontryagin, L.S., Boltyanskii, V.G., Gamkrelidze, R.V., Mishchenko, E.F.: Mathematical Theory of Optimal Processes, Gordon and Breach Science Publishers. 4 (1986)
20. Kirschner, D.E.: Using Mathematics to Understand HIV Immune Dynamics. Notices of the AMS. **43**, 191–200.115 (1996)
21. Fister, K.R., Lenhart, S., McNally, J.S.: Optimizing Chemotherapy in an HIV model. Elect. J. Diff. Equn. **32**, 1–12 (1998)
22. Culshaw, R.V., Rawn, S., Spiteri, R.J.: Optimal HIV treatment by maximising immuno response. J. Math. Biol. **48**, 545–562 (2004)
23. Joshi, H.R.: Optimal control of an HIV immunology model. Optimal Control Application and Methods. **23**, 199–213 (2002)
24. Culshaw, R.V., Rawn, S., Spiteri, R.J.: Optimal HIV treatment by maximising immunoresponse. J. Math. Biol. **48**, 545–562 (2004)
25. Chun,0 T. M., Fauci, A.S.: Latent reservoirs of HIV. Obstacles to the eradication of virus. Proc. Natl. Acad. Sci. USA. **96**, 10958–10961 (1999)

26. Adams, B.M., Banks, H.T., Kwon, H-D., Tran, H.T.: Dynamic Multidrug Therapies for HIV: Optimal and STI Control Approaches. Biosci. Eng. **1**, 223–242.4 (2004)
27. Ramratnam, B., Mittler, J.U.E., Zhang, L.Q., Boden, D., Hurley, A., Fang, F., Macken, C.A., Perelson, A.S., Markowitz, M. and Ho, D.D.: The decay of the latent reservoir of replication-competent HIV-1 is inversely correlated with the extent of residual viral replication during prolonged anti-retroviral therapy. Nat. Med. **6**, 82–85.127 (2000)
28. Gumel, A.B., Zhang, X.W., Shivkumar, P.N., Garba, M.L., Sahai, B.M.: A new mathematical model for assessing therapeutic strategies for HIV infection. J. Theo. Medicin. **4**(2), 147–155 (2002)
29. Roy, P.K., Chatterjee A.N.,: Effect of HAART on CTL Mediated Immune Cells: An Optimal Control Theoretic Approach. In: Sio long Ao, (ed.) Electrical Engineering and Applied Computing. New York: Len Gelman Springer. **90**, 595–607.132 (2011)
30. Marchettia, G., Meronia, L., Moltenia, G., Moronia, M., Clericib, M., Goria, A.: "Interleukin-2 immunotherapy exerts a differential effect on $CD4^+T$ and $CD8^+T$ cell dynamics". AIDS. **18**, 211–216 (2004)
31. Marchettia, G., Franzetti, F., Gori, A.: Partial immune reconstitution following highly active antiretroviral therapy: can adjuvant interleukin-2 fill the gap? J. Antimicrob. Chemother. **55**, 401–409.102 (2005)
32. Lakshmikantham, V., Bainov, D., Simeonov, P.S.: Theory of Impulsive Differential Equations. World Scientific (1989)
33. Bainov, D., Simeonov, Pavel, S.: Impulsive Differential Equations: Periodic Solutions and Applications, John Wiley and Sons, Incorporated (1993)
34. Lou, J., Chen, L., Ruggeri, T.: An impulsive differential model on post exposure prophylaxis to hiv-1 exposed individual. J. Biol. Sys. **17**(4), 659–683 (2009)
35. Lou, J., Smith, R.J.: Modelling the effects of adherence to the HIV fusion inhibitor enfuvirtide. J. Theor. Biol. **268**, 1–13 (2011)
36. Smith, R.J., Wahl, L.M.,: Drug resistance in an immunological model of HIV-1 infection with impulsive drug effects. Bull. Math. Biol. **67**(4), 783–813 (2005)
37. Smith, R.J.: Explicitly accounting for antiretroviral drug uptake in theoretical HIV models predicts long-term failure of protease-only therapy. J. Theor. Biol. V **251**(2) 227–237.152 (2008)
38. Smith, R.J., Aggarwala, B.D.: Can the viral reservoir of latently infected $CD4^+T$ cells be eradicated with antiretroviral HIV drugs? J. Math. Biol. **59**, 697–715.153 (2009)
39. Smith, R.J., Okano, J.T., Kahn, J.S., Bodine, E.N., Blower, S.: Evolutionary dynamics of complex networks of HIV drug-resistant strains: the case of San Francisco. Science. **327**(5966) 697–701.155 (2010)
40. Miron, R.E., Smith, R.J.: Modelling imperfect adherence to HIV induction therapy. BMC Infect. Dis. **10**(6) (2010)
41. Csajka, C., Verotta, D.: Pharmacokinetics pharmacodynamic modeling: history and perspectives. J. Pharmacokinet Pharmacodyn. **33**, 227–279 (2006)
42. Wagner, J.G.: A modern view of pharmacokinetics. J. Pharmacokinet Pharmacodyn. **1**, 363–401.166 (1973)

Chapter 5
Insight of Delay Dynamics

Abstract Delay effects during long-term HIV infection have been studied extensively. Here we have considered a time delay in the process of infection in the healthy T cells. Furthermore, we have also included a similar delay in the killing rate of infected CD4$^+$ T cells by Cytotoxic T-Lymphocytes (CTLs) and in the stimulation of CTLs. Since the process of generation of CTLs is not instantaneous, we have assimilated a realistic time lag in the production term of CTLs in our basic mathematical model of the HIV-1 infection. We have estimated the length of delay for which the stability of the system remains preserved. We have obtained the threshold value for delay parameter, below this critical value, interior equilibrium point becomes asymptotically stable. When the value of this delay parameter exceeds this threshold value, interior equilibrium point becomes unstable and a Hopf bifurcation occurs.

Keywords HIV-1 · CD4$^+$ T cells · Cytotoxic T-lymphocytes · Reverse transcriptase inhibitor · Time delay · Cell lysis

5.1 Delay Effect during Long-Term HIV Infection

Here we have considered a time delay in the process of infection in the healthy T cells. Further we also include a similar delay in the killing rate of infected CD4$^+$T cells by Cytotoxic T-Lymphocytes (CTLs) and in the stimulation of CTLs. In this section we consider the basic mathematical model proposed by [1].

In the basic model [1] including CTL response of an HIV infected system, the process of interaction between the infected CD4$^+$T cells and infectible CD4$^+$T cells through which transmission of the disease takes place, is considered as instantaneous. However the natural disease transmission process requires finite time. An infectible CD4$^+$T cell becomes the target of an already infected cell and is transformed to the infected class following a time consuming transmission process. This means that a time lag or delay exists in the process of disease transmission. Thus the model equation can be rewritten in the form,

© Springer Science+Business Media Singapore 2015

P.K. Roy, *Mathematical Models for Therapeutic Approaches to Control HIV Disease Transmission*, Industrial and Applied Mathematics, DOI 10.1007/978-981-287-852-6_5

$$\frac{dx}{dt} = \lambda - d_1 x - \beta_1 xy,$$

$$\frac{dy}{dt} = \beta_1 \int_{-\infty}^{t} x(u)y(u)F(t-u)du - d_2 y - \beta_2 yz,$$

$$\frac{dz}{dt} = sy - d_3 z. \tag{5.1}$$

Here we assume that the cells, which are produced infectious at time t are infected u time units ago, where u is distributed according to a probability distribution $F(u)$. It is known as the delay kernel and defined as $F(u) = \frac{\alpha^{n+1}u^n}{n!}e^{-\alpha u}$, where $\alpha > 0$ is constant and n is the order of delay [2]. Average delay is defined by $\tau = \int_0^\infty uF(u)du = \frac{n+1}{\alpha}$. Here the kernel takes the form $F(u) = \delta(u - \tau)$, where $\tau \geq 0$ is constant. Thus the system becomes the following delay differential equation with delay τ [3] as,

$$\frac{dx}{dt} = \lambda - d_1 x - \beta xy,$$

$$\frac{dy}{dt} = \beta x(t-\tau)y(t-\tau) - d_2 y - \beta_2 yz,$$

$$\frac{dz}{dt} = sy - d_3 z. \tag{5.2}$$

Here we also consider that the contract process between the uninfected and virus-producing cells is not instantaneous. Thus we include a delay, similar to the disease transmission term in the first equation in (5.2) and hence the set of equations take the form,

$$\frac{dx}{dt} = \lambda - d_1 x - \beta_1 x(t-\tau)y(t-\tau)P(t, \tau),$$

$$\frac{dy}{dt} = \beta_1 x(t-\tau)y(t-\tau)P(t, \tau) - d_2 y - \beta_2 yz,$$

$$\frac{dz}{dt} = sy - d_3 z, \tag{5.3}$$

with initial conditions $x(t - \tau) = 0$, $y(t - \tau) = 0$ for $t - \tau < 0$, $x(0) = x_0$, $y(0) = y_0$, and $z(0) = z_0$. Here $P(t, \tau)$ is the probability that a susceptible CD4$^+$T cell survives $[t - \tau, t)$. Let

$$P(\eta) = P(\text{susceptible x} - \text{cell survives } [\eta - \tau, \eta)) \tag{5.4}$$

with $P(0) = \exp(-d_1 \tau)$. In fact $P(\eta) = \exp(-d_1 \tau)$ for $\eta \leq \tau$. These equations are very complicated to analyze. If the time delay is relatively small, it will be approximated by the simpler set of equations:

$$\frac{dx}{dt} = \lambda - d_1 x - \beta_1 x(t - \tau)y(t - \tau),$$

$$\frac{dy}{dt} = \beta_1 x(t - \tau)y(t - \tau) - d_2 y - \beta_2 yz,$$

$$\frac{dz}{dt} = sy - d_3 z, \tag{5.5}$$

with the same initial conditions of (5.3).

5.1.1 Local Stability Analysis

The right-hand side of Eq. (5.5) is a smooth function of x, y, z (the variables) and the parameters, as long as the quantities are nonnegative, so local existence and uniqueness properties hold in the positive octant. The model equations (5.5) have equilibria $E_1(\frac{\lambda}{d}, 0, 0)$ and $E^*(x^*, y^*, z^*)$, where

$$x^* = \frac{(d_2 d_3 \beta_1 - d_1 s \beta_2) + \sqrt{(d_2 d_3 \beta_1 - d_1 s \beta_2)^2 + 4\beta_1^2 d_3 \beta_2 s \lambda}}{2 d_3 \beta_1^2}, \quad y^* = \frac{\lambda - d_1 x^*}{\beta_1 x^*} \text{ and } z^* = \frac{\beta_1 x^* - d_2}{\beta_2}.$$

Note that E^* is feasible and nontrivial if and only if $\frac{d_2}{\beta_1} < \frac{\lambda}{d_1}$.

Here we are interested to investigate the local stability of the interior equilibrium E^* of the delay-induced system (5.1).

Let $x'(t) = x(t) - x^*$, $y'(t) = y(t) - y^*$, $z'(t) = z(t) - z^*$ be the perturbed variables. The linearized form of the system (5.5) at $E^*(x^*, y^*, z^*)$ is given by,

$$\frac{dx'}{dt} = -d_1 x' - \beta_1 x'(t - \tau)y^* - \beta_1 y'(t - \tau)x^*,$$

$$\frac{dy'}{dt} = \beta_1 x'(t - \tau)y^* + \beta_1 y'(t - \tau)x^* - d_2 y' - \beta_2 y'z^* - \beta_2 z'y^*,$$

$$\frac{dz'}{dt} = sy' - d_3 z'. \tag{5.6}$$

The characteristic equation of system (5.6) is given by,

$$\rho^3 + (A + \beta_1 B e^{-\rho\tau})\rho^2 + (C + \beta_1 D e^{-\rho\tau})\rho + (E + \beta_1 F e^{-\rho\tau}) = 0, \tag{5.7}$$

where

$$A = d_1 + d_2 + d_3 + \beta_2 z^* \ (> 0), \quad B = y^* - x^*,$$
$$C = d_1(d_2 + d_3 + \beta_2 z^*) + d_2 d_3 + d_3 \beta_2 z^* + \beta_2 sy^* \ (> 0),$$
$$D = d_2 y^* + d_3 y^* + \beta_2 y^* z^* - d_1 x^* - d_3 x^*, \quad E = d_1(d_2 d_3 + d_3 \beta_2 z^* + \beta_2 sy^*),$$
$$F = y^*(\beta_2 sy^* + d_3 \beta_1 x^*) - d_1 d_3 x^*. \tag{5.8}$$

Now to determine the nature of the stability, we require the sign of the real parts of the roots of the system (5.7). Let

$$\Phi(\rho, \tau) = \rho^3 + (A + \beta_1 B e^{-\rho\tau})\rho^2 + (C + \beta_1 D e^{-\rho\tau})\rho + (E + \beta_1 F e^{-\rho\tau}).$$
(5.9)

For $\tau = 0$, i.e., for the nondelayed system

$$\Phi(\rho, 0) = \rho^3 + (A + \beta_1 B)\rho^2 + (C + \beta_1 D)\rho + (E + \beta_1 F) = 0.$$
(5.10)

Theorem 5.1.1 *For $\tau = 0$, the unique nontrivial equilibrium is locally asymptotically stable.*

Proof To show that the Routh–Hurwitz conditions [4] are satisfied, we need to show that $A + \beta_1 B > 0$, $E + \beta_1 F > 0$, and $(A + \beta_1 B)(C + \beta_1 D) > E + \beta_1 F$.

(i) $A + \beta_1 B = d_1 + d_2 + d_3 + \beta_2 z^* + \beta_1(y^* - x^*) = d_1 + d_3 + \beta_1 y^* > 0,$

(ii) $C + \beta_1 D = d_1(d_2 + d_3 + \beta_2 z^*) + d_2 d_3 + d_3 \beta_2 z^* + \beta_2 sy^* + \beta_1(d_2 y^* + d_3 y^* + \beta_2 y^* z^* - d_1 x^* - d_3 x^*) = d_1 d_3 + \beta_2 sy^* + \beta_1(d_3 y^* + \beta_1 x^* y^*) > 0,$

(iii) $E + \beta_1 F = d_1(d_2 d_3 + d_3 \beta_2 z^* + \beta_2 sy^*) + \beta_1 y^*(\beta_2 sy^* + d_3 \beta_1 x^*) - d_1 d_3 \beta_1 x^* = d_1 \beta_2 sy^* + \beta_1 y^*(\beta_2 sy^* + d_3 \beta_1 x^*) > 0.$

Using the above expressions, it is straightforward to show that Theorem 5.1.1 is true.

Substituting $\rho = u(\tau) + iv(\tau)$ in (5.9) and separating real and imaginary parts, we obtain the following transcendental equations:

$$u^3 - 3uv^2 + (u^2 - v^2)(A + \beta_1 B e^{-u\tau}\cos v\tau) + 2uv\beta_1 B e^{-u\tau}\sin v\tau + Cu$$
$$+ u\beta_1 D e^{-u\tau}\cos v\tau + v\beta_1 D e^{-u\tau}\sin v\tau + E + \beta_1 F e^{-u\tau}\cos v\tau = 0,$$
(5.11)

$$3u^2 v - v^3 + 2uvA + 2uv\beta_1 B e^{-u\tau}\cos v\tau - (u^2 - v^2)\beta_1 B e^{-u\tau}\sin v\tau + Cv$$
$$+ v\beta_1 D e^{-u\tau}\cos v\tau - \beta_1(uD + F)e^{-u\tau}\sin v\tau = 0.$$
(5.12)

5.1.2 Sufficient Conditions for Delay-Induced Instability

To find the conditions for nonexistence of delay-induced instability, we now use the following theorem [5]:

Theorem 5.1.2 *A set of necessary and sufficient conditions for the equilibrium E^* to be asymptotically stable for all $\tau \geq 0$ are the following:*

(i) *The real parts of all the roots of $\Phi(\rho, 0) = 0$ are negative,*

(ii) *For all real v and $\tau \geq 0$, $\Phi(iv, \tau) \neq 0$, where $i = \sqrt{-1}$.*

Proof Here $\Phi(\rho,0) = 0$ has roots, whose real parts are negative. Now for $v = 0$,

$$\Phi(0,\tau) = E + \beta_1 F \neq 0, \tag{5.13}$$

and for $v \neq 0$,

$$\Phi(iv,\tau) = -iv^3 - (A + \beta_1 Be^{-iv\tau})v^2 + iv(C + \beta_1 De^{-iv\tau}) + (E + \beta_1 Fe^{-iv\tau})$$
$$= 0. \tag{5.14}$$

Separating real and imaginary parts we get,

$$Av^2 - E = v\beta_1 D \sin v\tau + \beta_1(F - Bv^2)\cos v\tau, \tag{5.15}$$

$$-v^3 + Cv = -v\beta_1 D \cos v\tau + \beta_1(F - Bv^2)\sin v\tau. \tag{5.16}$$

Squaring and adding the above two equations, we get

$$(Av^2 - E)^2 + (-v^3 + Cv)^2 = v^2\beta_1^2 D^2 + \beta_1^2(F - Bv^2)^2. \tag{5.17}$$

Therefore, a sufficient condition for the nonexistence of a real number v satisfying $\Phi(iv,\tau) = 0$ can now be obtained from (5.17) as,

$$v^6 + (A^2 - 2C - \beta_1^2 B^2)v^4 + (C^2 - 2AE - \beta_1^2 D^2 + 2BF\beta_1^2)v^2 + E^2 - \beta_1^2 F^2 > 0, \tag{5.18}$$

for all real v. We can write this inequality in the form of

$$v^6 + Pv^4 + Qv^2 + R > 0, \tag{5.19}$$

where

$$P = A^2 - 2C - B^2\beta_1^2, \quad Q = C^2 - 2AE - D^2\beta_1^2 + 2BF\beta_1^2, \quad R = E^2 - \beta_1^2 F^2. \tag{5.20}$$

Therefore, conditions (*i*) and (*ii*) of Theorem 5.1.2 are satisfied if (5.18) holds.

5.1.3 Stability, Instability, and Bifurcation Results

Let us consider ρ and hence u and v as functions of τ. We are interested in the change of stability of E^*, which will occur at the values of τ for which $u = 0$ and $v \neq 0$. Let $\hat{\tau}$ be such that for which $u(\hat{\tau}) = 0$ and $v(\hat{\tau}) = \hat{v} \neq 0$. Then Eqs. (5.11) and (5.12) become,

$$- A\hat{v}^2 - \hat{v}^2\beta_1 B \cos \hat{v}\hat{\tau} + \hat{v}\beta_1 D \sin \hat{v}\hat{\tau} + \beta_1 F \cos \hat{v}\hat{\tau} + E = 0, \qquad (5.21)$$

$$- \hat{v}^3 + \hat{v}^2\beta_1 B \sin \hat{v}\hat{\tau} + C\hat{v} + \hat{v}\beta_1 D \cos \hat{v}\hat{\tau} - \beta_1 F \sin \hat{v}\hat{\tau} = 0. \qquad (5.22)$$

Now eliminating $\hat{\tau}$ we get,

$$\hat{v}^6 + (A^2 - 2C - \beta_1^2 B^2)\hat{v}^4 + (C^2 - \beta_1^2 D^2 - 2AE + 2BF\beta_1^2)\hat{v}^2$$
$$+(E^2 - \beta_1^2 F^2) = 0. \qquad (5.23)$$

To analyze the change in the behavior of the stability of E^* with respect to τ, we examine the sign of $\frac{du}{d\tau}$ as u crosses zero. If this derivative is positive (negative), then clearly a stabilization (destabilization) cannot take place at that value of $\hat{\tau}$. We differentiate equations (5.11) and (5.12) w.r.t. τ, then setting $\tau = \hat{\tau}$, $u = 0$, and $v = \hat{v}$ we get,

$$X\frac{du}{d\tau}(\hat{\tau}) + Y\frac{dv}{d\tau}(\hat{\tau}) = g, \quad -Y\frac{du}{d\tau}(\hat{\tau}) + X\frac{dv}{d\tau}(\hat{\tau}) = h, \qquad (5.24)$$

where

$$X = -3\hat{v}^2 + C + D\beta_1 \cos \hat{v}\hat{\tau} + 2\beta_1\hat{v}B \sin \hat{v}\hat{\tau} + \hat{\tau}[(\beta_1\hat{v}^2 B - \beta_1 F)\cos \hat{v}\hat{\tau} - \beta_1\hat{v}D \sin \hat{v}\hat{\tau}],$$
$$Y = -2A\hat{v} + \beta_1 D \sin \hat{v}\hat{\tau} - 2\hat{v}\beta_1 B \cos \hat{v}\hat{\tau} + \hat{\tau}[(\beta_1\hat{v}^2 B - \beta_1 F)\sin \hat{v}\hat{\tau} + \hat{v}\beta_1 D \cos \hat{v}\hat{\tau}],$$
$$g = -[(\beta_1\hat{v}^2 B - \beta_1 F)\sin \hat{v}\hat{\tau} + \hat{v}\beta_1 D \cos \hat{v}\hat{\tau}]\hat{v},$$
$$h = -[(\beta_1\hat{v}^2 B - \beta_1 F)\cos \hat{v}\hat{\tau} - \hat{v}\beta_1 D \sin \hat{v}\hat{\tau}]\hat{v}. \qquad (5.25)$$

Solving (5.24) we get,

$$\frac{du}{d\tau}(\hat{\tau}) = \frac{gX - hY}{X^2 + Y^2}. \qquad (5.26)$$

$\frac{du}{d\tau}(\hat{\tau})$ has the same sign as $gX - hY$.
After simplification and solving (5.21) and (5.22) we get,

$$gX - hY = \hat{v}^2[3\hat{v}^4 + 2(A^2 - 2C - B^2\beta_1^2)\hat{v}^2 + (C^2 - \beta_1^2 D^2 - 2AE + 2\beta_1^2 BF)]. \qquad (5.27)$$

Let $F(z) = z^3 + P_1 z^2 + P_2 z + P_3$, where

$$P_1 = A^2 - 2C - \beta_1^2 B^2, \quad P_2 = C^2 - \beta_1^2 D^2 - 2AE + 2\beta_1^2 BF, \quad P_3 = E^2 - \beta_1^2 F^2,$$

which is the left-hand side of Eq. (5.23) with $\hat{v}^2 = z$. Now

$$\frac{du}{d\tau}(\hat{\tau}) = \frac{\hat{v}^2}{X^2 + Y^2} \cdot \frac{dF}{dz}(\hat{v}^2). \qquad (5.28)$$

From the above results we can illustrate that if z is a positive simple root of $F(z) = 0$ and $z = \hat{v}^2$, then $\frac{du}{d\tau}(\hat{\tau}) \neq 0$. Also at $\tau = \hat{\tau}$, $u(\tau) = 0$, $v(\tau) = \hat{v}$, the root of the corresponding characteristic equation crosses the imaginary axis transversally. Hence in this case, according to Hopf bifurcation theorem [6], limit cycles arise and disappear at those values of $\hat{\tau}$. Hence $F(z)$ has zero, one, two or three positive real roots. We have the following theorem:

Theorem 5.1.3 (i) *If $F(z)$ has no positive real root, then the nontrivial equilibrium E^* is always locally asymptotically stable.*
(ii) *If $F(z)$ has one positive simple real root $\hat{v}$, then $F'(\hat{v}) > 0$. In this case there exists $\hat{\tau}_1 > 0$ such that the nontrivial equilibrium is locally asymptotically stable for $\tau \in [0, \hat{\tau}_1)$ and unstable for $\tau > \hat{\tau}_1$. $\hat{\tau}_1$ is the unique root in $(0, 2\pi/\hat{v}]$ of the equations:*

$$\sin \hat{v}\tau = \frac{(A\hat{v}^2 - E)\,\hat{v}D + (\hat{v}^3 - C\hat{v})(\hat{v}^2 B - F)}{\beta_1[(\hat{v}D)^2 + (\hat{v}^2 B - F)^2]} \quad and$$

$$\cos \hat{v}\tau = \frac{(\hat{v}^3 - C\hat{v})\,\hat{v}D - (A\hat{v}^2 - E)(\hat{v}^2 B - F)}{\beta_1[(\hat{v}D)^2 + (\hat{v}^2 B - F)^2]}. \tag{5.29}$$

As τ passes through $\hat{\tau}_1$, small amplitude periodic solutions arise by Hopf bifurcation.
(iii) *If $F(z)$ has two or three positive simple roots, then at the largest of these $F'(v) > 0$ and at the second largest $F'(v) < 0$. In this case there exists a sequence $\hat{\tau}_0 = 0$, $\hat{\tau}_1$, $\hat{\tau}_2$, $\hat{\tau}_3$, ... such that for $\tau \in (\hat{\tau}_{2j}, \hat{\tau}_{2j+1})$, $j = 0, 1, 2, 3, \ldots$ the nontrivial equilibrium is locally asymptotically stable, and for $\tau \in (\hat{\tau}_{2j+1}, \hat{\tau}_{2j+2})$, $j = 0, 1, 2, 3, \ldots$ the nontrivial equilibrium is unstable. $\hat{\tau}_{2j+1}$ is the next root of (5.29) after $\hat{\tau}_{2j}$ corresponding to $\hat{v}$, a positive root of $F(v) = 0$, where $F'(v) > 0$. $\hat{\tau}_{2j+2}$ is the next root of (5.29) after $\hat{\tau}_{2j+1}$ corresponding to a positive root of $F(v) = 0$, where $F'(v) < 0$. As τ passes through each τ_{2j+1}, small periodic solutions arise, which are caused by Hopf bifurcation and disappear by backward Hopf bifurcation as τ passes through $\hat{\tau}_{2j+2}$.*
(iv) *If $F(z)$ has a repeated positive real root, it is not possible to say as much.*

(a) Suppose that $\hat{\tau}_1$ is the first root of $\hat{\tau}$ given by (5.29) corresponding to a real positive root $\hat{v}$ of $F(v)$, which is associated with a simple root $\hat{v}$ with $F'(\hat{v}) > 0$. Then the nontrivial equilibrium is again locally stable for $\tau \in [0, \hat{\tau}_1)$ and unstable for $\tau \in (\hat{\tau}_1, \hat{\tau}^*)$, where $\hat{\tau}^*$ is the next value of $\hat{\tau}$ given by (5.29) after $\hat{\tau}_1$ corresponding to a real positive repeated root $\hat{v}$ of $F(v) = 0$. It is not possible to say anything about the system for $\tau \geq \tau^*$.
(b) Suppose that $\hat{\tau}_1$, the first value of $\hat{\tau}$ given by (5.29) corresponding to a positive real root $\hat{v}$ of $F(v) = 0$, corresponds to either a repeated root or a simple root with $F'(\hat{v}) < 0$. Then the nontrivial equilibrium is locally asymptotically stable for $\tau \in [0, \hat{\tau}_1^*)$, where $\hat{\tau}_1^*$ is the first value of $\hat{\tau}$ given by (5.29) corresponding to a repeated positive root $\hat{v}$ of $F(v) = 0$. Again it is not possible to say anything about the system for $\tau \geq \tau_1^*$.

This completes there analytical study of our basic mathematical model (5.5). In the next section, we shall explore this model and two other similar models numerically.

5.1.4 Numerical Simulations: Results and Discussions

In Fig. 5.1, we plot the time series solutions of the model variables corresponding to uninfected T cells, infected T cells, and CTL densities for different values of the delay factor τ. All the other model parameters are chosen to assume their standard values as in Table 3.2. The starting point is obtained by perturbing x from the non-trivial equilibrium value E^* given by $(x^*, y^*, z^*) = (212.5947, 18.5189, 185.1893)$. Figure 5.1a (Fig. 5.1, top left) has initial conditions $x(\theta) = 280.0$, $y(\theta) = 18.5189$ for $\theta \in (-\tau, 0]$ and $z(0) = 185.1893$. Figure 5.1b–d (Fig. 5.1, top right, bottom left, and bottom right respectively) has initial conditions $x(\theta) = 240.0$, $y(\theta) = 18.5189$ for $\theta \in (-\tau, 0]$ and $z(0) = 185.1893$. Figure 5.1a has $\tau = 1$ day, Fig. 5.1b has $\tau = 3$ days, Fig. 5.1c has $\tau = 6.2$ days, and Fig. 5.1d has $\tau = 8$ days. In this and the other

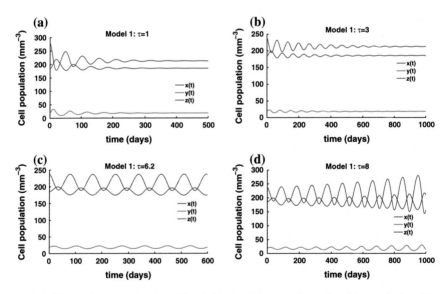

Fig. 5.1 Time series solutions of model variables for different values of τ with perturbation. The model parameters are as in Table 5.1 with $p = 0.001\,\text{mm}^{-3}\,\text{day}^{-1}$

Table 5.1 Parameters used in the models (5.5)

Parameter	λ	d_1	β_1	d_2	s	d_3	β_2	τ
Default value	10	0.01	0.002	0.24	0.2	0.02	0.001–0.005	1–15

All parameter values are taken from [1, 7, 8]

numerical simulations of our models which we shall present, the initial conditions for x and y have been taken to be constant in $(-\tau, 0]$ with x perturbed slightly from its nontrivial equilibrium value at E^*. We have done this because our main interest is the stability of the nontrivial equilibrium E^*.

We observe that introducing and then increasing the delay make the oscillations persist for longer. For the parameter values used $F(z)$ has one positive simple real root and the corresponding value of $\hat{\tau}_1$ is 6.2 days. Our simulations are consistent with the theoretical results. When $\tau = 0$ day, the nontrivial equilibrium E^* is locally asymptotically stable. Increasing the time delay τ makes the initial oscillations persist for longer, until we pass through the critical value $\hat{\tau}_1 = 6.2$ days (Fig. 5.1c), when there are regular cyclic oscillations. As τ increases beyond $\hat{\tau}_1$, the nontrivial equilibrium E^* becomes unstable and the oscillatory solutions increase in amplitude until they reach values of the susceptible and infected T cells and CTLs, where the approximate model becomes unrealistic.

Other simulations (not presented) indicate that as β_2 increases the amplitude of oscillations in the solutions diminishes and reduces the time span of the persistence of the solutions. The results seem to signify that there is a competition between the delay factor τ and the killing rate of infected T cells β_2 for dominance within the system.

Model equations including delay in the terms representing killing of virus-producing cells by CTLs and in the stimulation of CTLs:

In the basic delay model (5.5), we assume that killing of virus-producing cells by CTLs is an instantaneous process. But in reality there is a latency period during this process. Thus we consider a delay in the terms representing killing of virus-producing cells by CTLs and in the stimulation of CTLs. We thus consider solutions of the following model equations:

$$\frac{dx}{dt} = \lambda - d_1 x - \beta_1 xy,$$
$$\frac{dy}{dt} = \beta_1 xy - d_2 y - \beta_2 y(t - \tau)z,$$
$$\frac{dz}{dt} = sy(t - \tau) - d_3 z. \tag{5.30}$$

Again in a more general model, there would be a probability factor representing what happens to the activated virus-producing T cells between times $t - \tau$ and t, but we study the above approximation. One interpretation for this approximation is that it would be valid, when the time delay is small. Similar models have been studied by other authors, see for example [8–10]. A similar idea is also used in other contexts by [11, 12] and many other authors, see also our previous discussion. In Fig. 5.2, we again plot the time series solutions of the model variables corresponding to uninfected T cells, virus-producing T cells, and CTL densities with initial perturbation of the uninfected T cells from the nontrivial equilibrium value, changing the value of the delay factor τ. All of the other model parameters are chosen to assume their standard

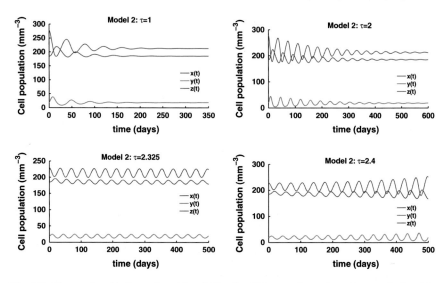

Fig. 5.2 Time series solutions of model variables for different values of τ with perturbation. The model parameters are as in Table 5.1 with $\beta_2 = 0.001 \, \text{mm}^{-3} \, \text{day}^{-1}$

values as in Table 5.1 where $\beta_2 = 0.001 \, \text{mm}^{-3} \, \text{day}^{-1}$. Figure 5.2a (top left) and b (top right) has the same initial conditions as Fig. 5.1a, Fig. 5.2c (bottom left) and d (bottom right) which have initial conditions $x(\theta) = 230.0$, $y(\theta) = 18.5189$ in $(-\tau, 0]$ and $z(0) = 185.1893$. Figure 5.2a has $\tau = 1$ day, Fig. 5.2b has $\tau = 2$ days, Fig. 5.2c has $\tau = 2.325$ days, and Fig. 5.2d has $\tau = 2.4$ days. Note that for both models our analytical results show that $\frac{a}{\beta_1} < \frac{\lambda}{d}$ is a necessary and sufficient condition for a unique equilibrium with virus . For $\tau = 0$ day, this equilibrium is locally asymptotically stable. We find that increase the delay enhances the oscillation in the solution and also makes the persistence of oscillation for longer time. In this respect, the results of the model are qualitatively similar to the results of the model with time delay in the infection term (5.5). The increasing oscillatory trends in Fig. 5.2d eventually reached regions of susceptible and virus-producing T cells and CTLs, where the approximate model is not valid. Increasing the time delay further do not lead to restabilization of the nontrivial equilibrium E^* at least for realistic values of the time delay.

Other simulations (not presented) are concentrated on the effect of changing the disease transmission term β_1. We find that as β_1 increases, the amplitude of oscillations reduces. Also as β_1 increases, the CTL population increases whereas the virus-producing and uninfected T cell populations decrease together.

5.1.5 Delay in Different Variants

In this model, we consider that delay exists in both (*i*) the process of infection of healthy T cells and (*ii*) the terms representing killing of virus-producing cells by CTLs and in the stimulation of CTLs together. We thus consider the following model equations:

$$\frac{dx}{dt} = \lambda - d_1 x - \beta_1 x(t - \tau_1) y(t - \tau_1),$$

$$\frac{dy}{dt} = \beta_1 x(t - \tau_1) y(t - \tau_1) - d_2 y - \beta_2 y(t - \tau_2) z,$$

$$\frac{dz}{dt} = sy(t - \tau_2) - d_3 z. \tag{5.31}$$

The results are shown in Fig. 5.3. Figure 5.3a–c (respectively top left, top right, bottom left) has the same initial conditions as Fig. 5.2c, d. Figure 5.3d (bottom right) has initial conditions as $x(\theta) = 225.0$, $y(\theta) = 18.5189$ in $(-\tau, 0]$ and $z(0) = 185.1893$. Figure 5.3a has $\tau_1 = 3$ days and $\tau_2 = 1$ day, Fig. 5.3b has $\tau_1 = 1$ day and $\tau_2 = 2$ days, Fig. 5.3c has $\tau_1 = 3$ days and $\tau_2 = 1.52$ days, and Fig. 5.3d has $\tau_1 = 4.1$ days and $\tau_2 = 1$ day. Again we note that for all three models, there is a unique nontrivial equilibrium if and only if $\frac{a}{\beta_1} < \frac{\lambda}{d}$ and in the third model if $\tau_1 = \tau_2 = 0$, then this equilibrium is locally asymptotically stable. We observe qualitatively that as either τ_1 or τ_2 increases, the long-term dynamics tends to become more oscillatory. Again for the more oscillatory solutions, the amplitude of the oscillations increases until they reach a region, where the approximated model is not valid. Increasing τ_1 and τ_2 further

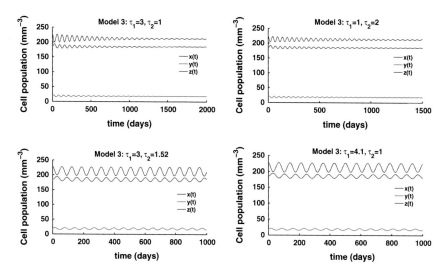

Fig. 5.3 Time series solutions of model variables for different values of τ with perturbation. The model parameters are as in Table 5.1. $\beta_2 = 0.001$ mm^{-3} day^{-1}

do not seem to lead to restabilization of the nontrivial equilibrium E^*. Figure 5.4 shows phase plots of the delay-induced system in the three different models: the model given by Eq. (5.5) (Fig. 5.1a, c, and d), the model given by Eq. (5.30) (Fig. 5.2a, c, and d), and the model given by Eq. (5.31) (Fig. 5.3a, c, and d). In Fig. 5.4a (top row left), d (middle row left), and g (bottom row left) the trajectories spiral inwards whereas in Fig. 5.4c (top row right) and f (middle row right) the trajectories spiral outwards. In the top row of Fig. 5.4, we see that there is a critical threshold value $\tau_{th} = 6.2$ days such that the system (5.5) is locally asymptotically stable for $\tau < \tau_{th}$ and unstable for $\tau > \tau_{th}$. These numerical results match our analytical predictions exactly. In the middle row of Fig. 5.4, we see that the system (5.30) moves from local asymptotic stability to increasing oscillatory solutions, as τ passes through the approximate threshold value $\tau_{th} = 2.325$ days. The bottom row of Fig. 5.4 suggests that the system (5.31) moves from being locally asymptotically stable for τ_1 and τ_2 small to oscillatory solutions as τ_1 and τ_2 become larger. Again as we increase τ_1 and τ_2 further, we observe oscillatory solutions of increasing amplitude diverging

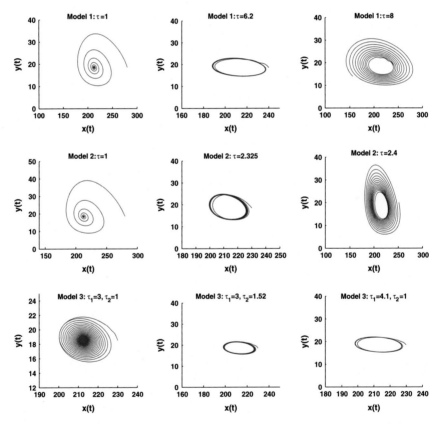

Fig. 5.4 Phase plots of model variables for different values of τ with perturbation. The model parameters are as in Table 5.1 with $\beta_2 = 0.001\,\mathrm{mm}^{-3}\,\mathrm{day}^{-1}$

from the nontrivial equilibrium E^*. Thus the conclusions of our numerical analysis can be briefly summarized by saying that increasing the delay in the system causes increasing oscillations and instability in all the models which we have considered.

5.2 Delay-Induced System in Presence of Cure Rate

In this section, we introduce a time delay into Eq. (2.4) on the assumption that the transmission of the disease is not an instantaneous process. In reality, there is a time lag between the process of infection of cells to cells becoming actively infected. Thus the model becomes,

$$
\begin{aligned}
\dot{x} &= \lambda - d_1 x - \beta_1 x(t-\tau)y(t-\tau) + \delta y, \\
\dot{y} &= \beta_1 x(t-\tau)y(t-\tau) - (d_2 + \delta)y - \beta_2 yz, \\
\dot{z} &= sy - d_3 z,
\end{aligned}
\tag{5.32}
$$

with the initial conditions

$$
x(\theta) = x_\theta, \ y(\theta) = y_\theta, \ z(\theta) = z_\theta, \ \theta \in [-\tau, 0].
\tag{5.33}
$$

5.2.1 Analysis

In this section, we also study the local and global stability of the disease-free equilibrium E_0 of the delay differential equations. Here we consider the two stability function, when $R_0 < 1$ and $R_0 > 1$. For delayed system, we also find two steady states $E_0 = (x_0, 0, 0)$ and $\bar{E} = (\bar{x}, \bar{y}, \bar{z})$ similar to the nondelayed system. In previous section, we have seen that for $R_0 > 1$, E_0 is unstable. The result is same as before for delayed system. Then E_0 is unstable, if $R_0 > 1$ and the system moves towards infected steady state $\bar{E}$. To study the stability of the steady state $\bar{E}$, we linearize the system by substituting $X(t) = x(t) - \bar{x}$, $Y(t) = y(t) - \bar{y}$, $Z(t) = z(t) - \bar{z}$. Then the linearized system of the equation (5.32) at $\bar{E}$ is given by,

$$
\begin{aligned}
\dot{X} &= -d_1 X - \beta_1 \bar{y} X(t-\tau) - \beta_1 \bar{x} Y(t-\tau) + \delta Y, \\
\dot{Y} &= \beta_1 \bar{y} X(t-\tau) + \beta_1 \bar{x} Y(t-\tau) - (d_2 + \delta)Y - \beta_2 \bar{z} Y - \beta_2 \bar{y} Z, \\
\dot{Z} &= sY - d_3 Z.
\end{aligned}
\tag{5.34}
$$

The characteristic equation of the system (5.34) is given by,

$$
\rho^3 + a_1 \rho^2 + a_2 \rho + a_3 + (a_4 \rho^2 + a_5 \rho + a_6)e^{-\rho\tau} = 0,
\tag{5.35}
$$

where

$$a_1 = d_1 + d_2 + d_3 + \delta + \beta_2 \bar{z}, \quad a_2 = (d_1 + d_3)(d_2 + \delta + \beta_2 \bar{z}) + d_3(d_1 + \beta_2 s \bar{y}),$$
$$a_3 = d_1 d_3(d_2 + \delta + \beta_2 \bar{z}) + \beta_2 d_1 s \bar{y}, \quad a_4 = \beta_1(\bar{y} - \bar{x}),$$
$$a_5 = \beta_1 \{\bar{y}(d_2 + d_3 + \beta_2 \bar{z}) - \bar{x}(d_1 + d_3)\}, \quad a_6 = \beta_1 \{\bar{y}(d_2 d_3 + \beta_2 d_3 \bar{z} + \beta_2 s \bar{y}) - d_1 d_2 \bar{x}\}.$$
$$(5.36)$$

We know that the infected steady state is asymptotically stable, if all roots of the characteristic equation have negative real parts. Since Eq. (5.35) is a transcendental equation and it has infinitely many eigenvalues, it is very difficult to deal with this equation. From classical Routh–Hurwitz criterion, we cannot discuss the characteristic equation. By Rouche's theorem and continuity in τ, the characteristic equation (5.35) has roots with positive real parts if and only if it has purely imaginary roots, then we can find out the condition for all eigenvalues to have negative real parts. Let $\rho = u(\tau) + iv(\tau), \quad (v > 0)$ be the eigenvalue of the characteristic equation (5.35), where u and v depend on τ. For nondelayed system ($\tau = 0$), $\bar{E}$ is stable if $u(0) < 0$. Since τ is continuous, then for small value of $\tau > 0$, we still have $u(\tau) < 0$ and $\bar{E}$ remains stable. If for certain $\tau_0 > 0$, $u(\tau_0) = 0$, then the steady state $\bar{E}$ losses its stability and it becomes unstable, when $u(\tau) > 0$ for $\tau > \tau_0$. Also if all the roots of the characteristic equation (5.35) stand real (i.e., $v(\tau) = 0$), then $\bar{E}$ is always stable. Now to see whether Eq. (5.35) has purely imaginary root or not, we put $\rho = iv$ in Eq. (5.35) and separating the real and imaginary parts, we have,

$$a_1 v^2 - a_3 = (a_6 - a_4 v^2)\cos(v\tau) + a_5 v \sin(v\tau) \quad \text{and} \quad (5.37)$$
$$v^3 - a_2 v = a_5 v \cos(v\tau) - (a_6 - a_4 v^2)\sin(v\tau). \quad (5.38)$$

Squaring and adding the above equations we get,

$$v^6 + (a_1^2 - 2a_2 - a_4^2)v^4 + (a_2^2 - a_5^2 + 2a_4 a_6 - 2a_1 a_3)v^2 + (a_3^2 - a_6^2) = 0. \quad (5.39)$$

Let $\xi = v^2, \quad \alpha_1 = a_1^2 - 2a_2 - a_4^2, \quad \alpha_2 = a_2^2 - a_5^2 + 2a_4 a_6 - 2a_1 a_3, \alpha_3 = a_3^2 - a_6^2.$
Then Eq. (5.39) becomes,

$$F(\xi) = \xi^3 + \alpha_1 \xi^2 + \alpha_2 \xi + \alpha_3 = 0. \quad (5.40)$$

Since $\alpha_3 = a_3^2 - a_6^2 > 0$ and $\alpha_2 > 0$, then Eq. (5.40) has no positive real root. Now,

$$F'(\xi) = 3\xi^2 + 2\alpha_1 \xi + \alpha_2 = 0. \quad (5.41)$$

Then the roots of the equation (5.41) are,

$$\xi_1 = \frac{-\alpha_1 + \sqrt{\alpha_1^2 - 3\alpha_2}}{3} \quad \text{and} \quad \xi_2 = \frac{-\alpha_1 - \sqrt{\alpha_1^2 - 3\alpha_2}}{3}.$$

Since $\alpha_2 > 0$, then $\sqrt{\alpha_1^2 - 3\alpha_2} < \alpha_1$.

Hence both the roots are negative. Thus Eq. (5.41) does not have any positive root. Since $F(0) = \alpha_3 \geq 0$, Eq. (5.39) has no positive root. In brief, we can get the theorem stated below.

Theorem 5.2.1 *If the system satisfies*

(i) $R_0 > 1$,
(ii) $A_1 A_2 - A_3 > 0$, *and*
(iii) $\alpha_3 \geq 0$ *and* $\alpha_2 > 0$, *then the infected steady state is asymptotically stable for all* $\tau \geq 0$.

Now if the above conditions are not satisfied, i.e., if (i) $\alpha_3 < 0$, then $F(0) < 0$ and $\lim_{\xi \to \infty} F(\xi) = \infty$. Thus Eq. (5.35) has at least one positive root say ξ_0, then Eq. (5.40) has at least one positive root, say v_0. If $\alpha_2 < 0$, then $\sqrt{\alpha_1{}^2 - 3\alpha_2} > \alpha_1$, then from (5.42), $\xi_1 = \frac{-\alpha_1 + \sqrt{\alpha_1{}^2 - 3\alpha_2}}{3} > 0$. Hence the positive root v_0 exists for (5.39). Thus the characteristic equation (5.39) has a pair of purely imaginary roots $\pm i v_0$. From Eqs. (5.36) and (5.37) we have,

$$\tau_n = \frac{1}{v_0} \arccos \frac{a_5 v_0 (v_0{}^3 - a_2 v_0) - (a_3 - a_1 v_0{}^2)(a_6 - a_4 v_0{}^2)}{a_5{}^2 v_0{}^2 + (a_6 - a_4 v_0{}^2)^2} + \frac{2j\Pi}{v_0}, j = 1, 2, 3 \ldots$$

To show that at $\tau = \tau_0$, there exists a Hopf bifurcation, we need to verify the transversal condition $\frac{d}{d\tau}(Re\rho(\tau))|_{\tau=\tau_0} > 0$. By differentiation (5.35) with respect to τ, we get,

$$(3\rho^2 + 2a_1\rho + a_2)\frac{d\rho}{d\tau} + [e^{-\rho\tau}(2a_4\rho + a_5) - \tau e^{-\rho\tau}(a_4\rho^2 + a_5\rho + a_6)]\frac{d\rho}{d\tau}$$
$$= \rho e^{-\rho\tau}(a_4\rho^2 + a_5\rho + a_6).$$

$$(5.42)$$

From (5.42) we get,

$$\sin\{\frac{d(Re\rho)}{d\tau}\}|_{\rho=iv_0} = \sin\{Re(\frac{d\rho}{d\tau})^{-1}\}.$$

Since v_0 is the largest positive root of the equation (5.39), then we have $\frac{d}{d\tau}(Re\rho(\tau))|_{\tau=\tau_0} > 0$. So when $\tau > \tau_0$, the real part of $\rho(\tau)$ becomes positive and thus the system $\bar{E}$ becomes unstable. Thus we have the theorem.

Theorem 5.2.2 *If (i)* $R_0 > 1$, *(ii)* $A_1 A_2 - A_3 > 0$, *(iii)* $\alpha_3 \leq 0$, $\alpha_2 < 0$, *and* $\alpha_2 \geq 0$, *then the infected steady state is asymptotically stable for all* $\tau < \tau_0$ *and* $\bar{E}$ *becomes unstable, when* $\tau > \tau_0$, *where*

$$\tau_0 = \frac{1}{v_0} \arccos \frac{(a_3 - a_1 v_0{}^2)(a_6 - a_4 v_0{}^2) - a_5 v_0 (v_0{}^3 - a_2 v_0)}{v_0{}^2 a_5{}^2 + (a_6 - a_4 v_0{}^2)^2}.$$

Hence at $\tau = \tau_0$, Hopf bifurcation occurs, i.e., a family of periodic solution bifurcates for $\bar{E}$ as τ passes through its critical value τ_0. Thus we can comment from the above Theorem 5.2.2 that the delay model reveals Hopf bifurcation at a certain value τ_0, if the parameter satisfies the conditions (*i*), (*ii*) *and* (*iii*). Thus the delay in the disease transmission makes the system stable for the condition stated in Theorem 5.2.1 and the system moves towards unstable region for the condition established in Theorem 5.2.2.

5.2.2 Numerical Simulation

Figure 5.5 shows that for nondelayed system, as the cure rate is improved, the uninfected T cell population increases, where as the infected T cell and CTL population decrease. Here we plot the trajectories for $\tau = 0$, $\tau = 1$, and $\tau = 5$. This figure shows that as delay is introduced, the system oscillates and as delay is increased,

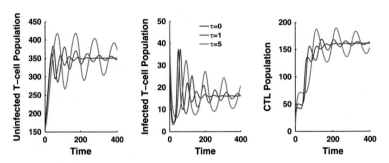

Fig. 5.5 Time series solution of the model variables for delayed and nondelayed systems, keeping all other parameters as in Table 5.2

Table 5.2 Parameters used in the model (5.32)

Parameter	Definition	Default value assigned (day^{-1})
λ	Constant rate of production of CD4$^+$T cells	$10.0\,\text{mm}^{-3}$
d_1	Death rate of uninfected CD4$^+$T cells	0.01
β_1	Rate of contact between x and y	$0.002\,\text{mm}^{-3}$
d_2	Death rate of virus-producing cells	0.24
β_2	Killing rate of virus-producing cells by *CTLs*	$0.001\,\text{mm}^{-3}$
s	Rate of simulation of *CTLs*	0.2
d_3	Death rate of *CTLs*	0.02
δ	Rate of cure	0.02

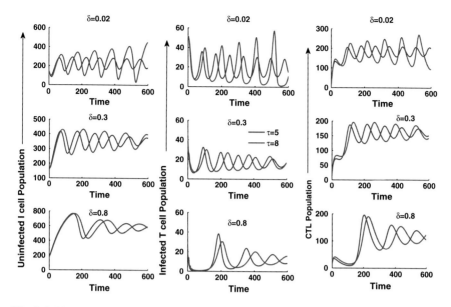

Fig. 5.6 The system behavior for different values of δ and τ, keeping all other parameters as in Table 5.2

the amplitude of oscillation increases. Thus incorporation of delay makes the system unstable.

Figure 5.6 represents the time series solution for different values of δ and τ. In this figure, we see that as delay increases from $\tau = 5$ to $\tau = 8$, the amplitude of oscillation of the solution trajectories is increased. If δ is increased from $\delta = 0.02$ to $\delta = 0.8$, the effect of delay is controlled. Thus if the cure rate is improved, the delay effect can be controlled. Thus the stability of the system is preserved.

5.3 Delay Effect during Early Stage of Infection

In this section, we have considered different time delayed models of HIV infection in human immune system including its response to RTI drug therapies. The discrete time delay is incorporated in the processes of infecting healthy T cells. We have analyzed the dynamics of the models in different cases to understand how the HIV infected immune system responds to varying levels of drugs applied under the systematic therapeutic procedure. The aim is to study and compare the dynamics of the proposed model including delay in different cases to explore the crucial system parameters and their ranges. It will be done in order to obtain different theoretical behaviors predicted from the interaction between targeted and infected $CD4^+T$ cells and also CTL responses against virus-producing cells. We have considered a time delay in the process of infection in the targeted T cells during the early stage of infection in the presence of proliferation process of targeted T cells.

5.3.1 General Mathematical Model

In the model (3.1), it is assumed that in an HIV infected system including CTL response, the process of interaction between infected and uninfected CD4$^+$T cells through which the transmission of the disease takes place is instantaneous. However in reality, there is a time delay between these two events.

(i) the first effective contact between infected and uninfected cells and
(ii) the target cells become effectively infectious.

Under the suitable conditions, here we assume a time delay $\tau \geq 0$ and the model equation is given as follows:

$$\dot{x} = \lambda + px(1 - \frac{x}{T_m}) - d_1 x - \beta_1 xy,$$
$$\dot{y} = \beta_1 x(t - \tau)y(t - \tau) - d_2 y - \beta_2 yz,$$
$$\dot{z} = sy - d_3 z, \tag{5.43}$$

under the initial conditions, $x(0) \geq 0$, $y(0) \geq 0$, $z(0) \geq 0$.

5.3.1.1 Theoretical Analysis: Local Stability and Delay-Induced Instability

The right-hand side of Eq. (5.43) is a smooth function of the variables x, y, and z and the parameters are all nonnegative. Then the model equation (5.43) has the following equilibria of all the coordinate planes:

(i) an uninfected steady state $E_0(x_0, 0, 0)$ and
(ii) an infected steady state $E^*(x^*, y^*, z^*)$,

where

$$x_0 = \frac{T_m}{2p}[(p - d_1) + \sqrt{(p - d_1)^2 + 4\frac{p\lambda}{T_m}}],$$

$$x^* = \frac{(\frac{d_2 d_3 \beta_1}{s\beta_2} + p - d_1) + \sqrt{(\frac{d_2 d_3 \beta_1}{s\beta_2} + p - d_1)^2 + 4\lambda(\frac{d_3 \beta_1^2}{s\beta_2} + \frac{p}{T_m})}}{2(\frac{d_3 \beta_1^2}{s\beta_2} + \frac{p}{T_m})}, \quad y^* = \frac{d_3}{s\beta_2}(\beta_1 x^* - d_2) \text{ and } z^* = \frac{\beta_1 x^* - d_2}{k},$$

satisfying the following inequality,

$$x^* > \frac{d_2}{\beta_1} \quad \text{and} \quad p > (d_1 - \frac{d_2 d_3 \beta_1}{s\beta_2}). \tag{5.44}$$

Here p is the bifurcation parameter. Now E^* exists, whenever $x^* < x_0$. We get a critical value of proliferation rate, i.e., $p_{crit} = \frac{d_1 x_0 - \lambda}{p(1 - \frac{x_0}{T_m})}$.

Remark 5.3.1 Whenever $p < p_{crit}$, E_0 is stable and E^* does not exist. For $p > p_{crit}$, E_0 becomes unstable and E^* exists. When $p = p_{crit}$, E_0 and E^* collide and there exists a transcritical bifurcation.

Here we are interested to investigate the local stability around the interior equilibrium E^* of the delay-induced model (5.43). To undergo the stability of the steady state E^*, let $X(t) = x(t) - x^*$, $Y(t) = y(t) - y^*$, $Z(t) = z(t) - z^*$. Then the linearized form of the system (5.43) at E^* is given by,

$$\dot{X} = -(d_1 + \frac{2px^*}{T_m} + \beta_1 y^* - p)X - \beta_1 x^* Y,$$
$$\dot{Y} = \beta_1 X(t - \tau)y^* + \beta_1 Y(t - \tau)x^* - (d_2 + \beta_2 z^*)Y - \beta_2 y^* Z,$$
$$\dot{Z} = sY - d_3 Z. \tag{5.45}$$

The characteristic equation of the system (5.45) is given by,

$$\rho^3 + (M + a_1)\rho^2 + (a_3 + Ma_1)\rho + Ma_3 + [a_2\rho^2 + (d_2a_2 + a_4)\rho + d_3a_4]e^{-\rho\tau} = 0, \tag{5.46}$$

where

$$a_1 = d_2 + d_3 + \beta_2 z^*, \quad a_2 = -\beta_1 x^*, \quad a_3 = d_2 d_3 + d_3\beta_2 z^* + s\beta_2 y^*,$$
$$a_4 = \beta_1^2 x^* y^* - M\beta_1 x^*, \quad M = d_1 + \frac{2px^*}{T_m} + \beta_1 y^* - p. \tag{5.47}$$

If the roots of the characteristic equation have negative real parts, then E^* is asymptotically stable. Let,

$$\psi(\rho, \tau) = \rho^3 + (M + a_1)\rho^2 + (a_3 + Ma_1)\rho + Ma_3$$
$$+ [a_2\rho^2 + (d_3a_2 + a_4)\rho + d_3a_4]e^{-\rho\tau} = 0. \tag{5.48}$$

For $\tau = 0$,

$$\psi(\rho, 0) = \rho^3 + (M + a_1 + a_2)\rho^2 + \{Ma_1 + a_3 + (d_3a_2 + a_4)\}\rho + (Ma_3 + d_3a_4)$$
$$= 0. \tag{5.49}$$

According to the Routh–Hurwitz criterion [5], the nondelayed system (5.49) has all eigenvalues with negative real parts if and only if,

$$M + a_1 + a_2 > 0, \quad Ma_3 + d_3a_4 > 0,$$
$$\{Ma_1 + a_3 + (d_3a_2 + a_4)\}(M + a_1 + a_2) - (Ma_3 + d_3a_4) > 0. \tag{5.50}$$

Therefore, we get the proposition.

Proposition 5.3.1 *The infected steady state E^* for non-delayed system is asymptotically stable, if the inequality (5.50) together with (5.44) is satisfied.*

Now if $\tau > 0$, then the characteristic equation (5.48) is a transcendental equation and it has infinitely many eigenvalues. So the Routh–Hurwitz criterion cannot be used in this equation. Here we try to find out the condition of stability for infected steady state with a finite delay.

Let $\rho(\tau) = u(\tau) + iv(\tau)$, $(v > 0)$. Since the equilibrium point E^* has been stable, it follows that when $\tau = 0, u(0) < 0$. Now if $\tau > 0, u(\tau) < 0$, then E^* is still stable. If $u(\tau_0) = 0$ for certain $\tau_0 > 0$, then the steady state E^* loses its stability and it becomes unstable, when $u(\tau)$ becomes positive. If $v(\tau_0)$ does not exist, then the characteristic equation (5.48) does not have any purely imaginary root for all delay. Hence the steady state E^* is always stable. We shall show that the above statement is true for the characteristic equation (5.48). Suppose, $\rho = iv(\tau)$ is a root of the equation (5.48) if and only if,

$$- iv^3 - (M + a_1)v^2 + iv(a_3 + Ma_1) + Ma_3$$
$$+ [-a_2v^2 + iv(d_3a_2 + a_4) + d_3a_4](\cos v\tau - i \sin v\tau) = 0. \qquad (5.51)$$

Separating real and imaginary parts, we obtain the following transcendental equations:

$$Ma_3 - v^2(M + a_1) = -(d_3a_4 - a_2v^2) \cos v\tau - v(d_3a_2 + a_4) \sin v\tau, \qquad (5.52)$$

$$- v^3 + v(a_3 + Ma_1) = -v(d_3a_2 + a_4) \cos v\tau + (d_3a_4 - a_2v^2) \sin v\tau. \qquad (5.53)$$

Squaring and adding the above two equations we get,

$$v^6 + [(M + a_1)^2 - a_2^2 - 2(a_3 + Ma_1)]v^4 + [(a_3 + Ma_1)^2$$
$$-2Ma_3(M + a_1) + 2d_3a_2a_4 - (d_3a_2 + a_4)^2]v^2 + M^2a_3^2 - d^2{}_3a_4^2 = 0. \qquad (5.54)$$

Let $\xi = v^2$,
$P_1 = (M + a_1)^2 - a_2^2 - 2(a_3 + Ma_1)$,
$P_2 = (a_3 + Ma_1)^2 - 2Ma_3(M + a_1) + 2d_3a_2a_4 - (d_3a_2 + a_4)^2$,
$P_3 = M^2a_3^2 - d^2{}_3a_4^2$.

Then Eq. (5.54) becomes,

$$F(\xi) = \xi^3 + P_1\xi^2 + P_2\xi + P_3 = 0. \qquad (5.55)$$

Since $P_3 = M^2a_3^2 - d^2{}_3a_4^2 > 0$ and $P_2 > 0$, Eq. (5.55) has no positive real root. Now,

$$\frac{dF(\xi)}{d\xi} = 3\xi^2 + 2P_1\xi + P_2 = 0. \qquad (5.56)$$

The roots of the equation (5.56) are,

$$\xi_1 = \frac{-P_1 + \sqrt{P_1^2 - 3P_2}}{3} \quad and \quad \xi_2 = \frac{-P_1 - \sqrt{P_1^2 - 3P_2}}{3}. \tag{5.57}$$

Since $P_2 > 0$, then $\sqrt{P_1^2 - 3P_2} < P_1$. Hence neither ξ_1 nor ξ_2 is positive. Thus Eq. (5.56) does not have any positive root. Since $F(0) = P_3 \geq 0$, the equation has no positive root.

Assumption 5.2.1 If $P_3 \geq 0$ and $P_2 > 0$, we can claim that there exists no v such that iv is the eigenvalue of the characteristic equation (5.48) is negative for all delay $\tau > 0$.

Proposition 5.3.2 *If the system satisfies*

1. $\{a_3 + Ma_1 + d_3(a_2 + a_3)\}(M + a_1 + a_2) - (Ma_3 + d_3a_4) > 0$ *and*
2. $P_3 \geq 0$ *and* $P_2 > 0$,

then the infected steady state E^* is asymptotically stable for all $\tau > 0$.

For Proposition 5.3.2, we can say that if the parameters in Table 5.3 satisfy the conditions (*i*) and (*ii*), obviously the infected steady state of the delay model is asymptotically stable for all delay values. Thus we can say from the above condition that the system is stable and it is independent of the delay.

Assumption 5.2.2 If $P_3 < 0$ and $P_2 < 0$, then there exists positive root v_0 such that characteristic equation has a conjugate pair of purely imaginary roots $\pm iv_0$. For Eq. (5.55), $F(0) < 0$ and $\lim_{\xi \to \infty} F(\xi) = \infty$. This Eq. (5.55) has at least one positive root denoted by ξ_0. Now if $P_2 < 0$, then $\sqrt{P_1^2 - 3P_2} > P_1$, and $\xi_1 = \frac{-P_1 + \sqrt{P_1^2 - 3P_2}}{3} > 0$. This implies that Eq. (5.54) has a pair of purely imaginary roots $\pm iv_0$. Now for $\tau = \tau_0$, $u(\tau_0) = 0$ and $v(\tau_0) = v_0$. Then from Eqs. (5.52) and (5.53) we get,

$$\tau_n = \frac{1}{v_0} \arctan[\frac{v_0(t_1t_2 + t_3t_4)}{t_1t_3 - v_0^2t_2t_4}] + \frac{2n\pi}{v_0}, \quad n = 0, 1, 2, \ldots, \tag{5.58}$$

where $v_0^2 = \frac{-P_1 + \sqrt{P_1^2 - 3P_2}}{3}$ and $t_1 = d_3a_4 - a_2v_0^2$, $t_2 = v_0^2 - a_3 - Ma_1$, $t_3 = M(a_3 - v_0^2) - v_0^2a_1$, $t_4 = (d_3a_2 + a_4)$.

Note that the following transversality condition, $\frac{d}{d\tau}Re\lambda(\tau) \|_{\tau=\tau_0} = \frac{d}{d\tau}(u(\tau)) \|_{\tau=\tau_0} > 0$. So the real part of $\rho(\tau)$ becomes positive, when $\tau > \tau_0$ and the steady state becomes unstable.

Table 5.3 Parameters used in the model (5.43)

Parameter	λ	p	T_m	d_1	β_1	d_2	s	d_3	β_2	β_1'
Default value	10	0.03	1500	0.02	0.002	0.24	0.002	0.2	0.02	0.0008

All parameter values are taken from [1, 7, 8]

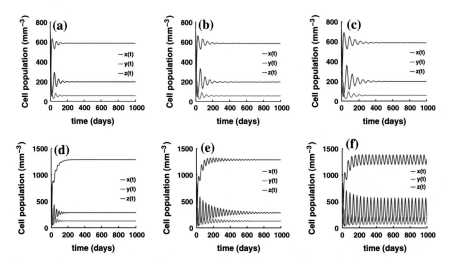

Fig. 5.7 Time series solution of the delayed system with different values of p and τ. Other parameters remain unchanged as in Table 5.3

Proposition 5.3.3 *If*

(i) $a_3 + Ma_1 + (d_3a_2 + a_4)(M + a_1 + a_2) - (Ma_3 + d_3a_4) > 0$ *and*
(ii) $P_3 < 0$ *or* $P_3 \geq 0$ *and* $P_2 < 0$,

then E^* is asymptotically stable, when $\tau < \tau_0$ and unstable, when $\tau > \tau_0$, where

$$\tau_0 = \frac{1}{v_0} \arctan[\frac{v_0(t_1t_2 + t_3t_4)}{t_1t_3 - v_0^2t_2t_4}], \quad n = 0. \tag{5.59}$$

Thus when $\tau = \tau_0$, Hopf bifurcation occurs, i.e., a family of periodic solution bifurcates for E^* as τ passes through the critical values τ_0.

5.3.2 Numerical Simulation

We now numerically illustrate the change of the stability due to varying the time delay. We choose the initial conditions of the parameters given as in Table 5.3. At $t = 0$, the values of the model variables are considered as $x(0) = 100$, $y(0) = 50$, $z(0) = 2$ and units are mm^{-3}. It should be noted that the asymptotic time series solutions of the model equation do not depend on the choice of the initial values of the model variables. A variation of the parameter k is restricted by the condition $\frac{ks}{d_3} \sim 0.01 - 0.05$. The parameters s and d_3 are as mentioned in Table 5.3.

Figure 5.7 represents the time series solutions of the model variables corresponding to uninfected T cells, infected T cells, and CTL densities for different

values of the delay factor τ and p, keeping $\beta_2 = 0.0001$. All the other model parameters are assumed from their standard values as in Table 5.3. The starting point
is obtained by perturbing x from the nontrivial equilibrium value E^*, given by
$(x^*, y^*, z^*) = (730, 25, 245)$. Figure 5.7a, and d have $\tau = 1$ day, Fig. 5.7b and e have
$\tau = 3$ days, Fig. 5.7c and f have $\tau = 5$ days. In the top panel of Fig. 5.7a–c, we keep
$p = 0.1$. From these three figures, we have observed that with increasing delay, the
amplitude of oscillation increases. But as time progresses, the system moves to its
infected equilibrium. In the bottom panel of Fig. 5.7d–f, we keep $p = 0.3$. From
these figures, we have observed that when delay increases from $\tau = 3$ to $\tau = 5$, the
system exhibits a regular cyclic oscillation. We have observed that increasing delay
makes the oscillations persist for longer. For the parameter values used in $F(z)$ has
one positive simple real root and the corresponding value of $\hat{\tau}_1$ is 3.7709 days. Our
numerical simulations are reliable with the theoretical outcomes. When $\tau = 0$ day,
the nontrivial equilibrium E^* is locally asymptotically stable. Increasing the time
delay τ makes the initial oscillations persist for longer, until we pass through the
critical value $\hat{\tau}_1 = 3.7709$ days (Fig. 5.7f), when there are regular cyclic oscillations.
As τ increases beyond $\hat{\tau}_1$, the nontrivial equilibrium E^* becomes unstable and the
oscillatory solutions increase in amplitude until they reach the values of susceptible
and infected T cells and CTLs, where the approximate model becomes unrealistic.

Figure 5.8 represents the time series solutions of the model variables corresponding to uninfected T cells, infected T cells, and CTL densities for different values
of the delay factor τ and β_2, keeping $p = 0.3$. Figure 5.7a and c have $\tau = 1$ day,
Fig. 5.7b and d have $\tau = 5$ days. In the top panel of Fig. 5.7a and b, we keep
$\beta_2 = 0.0005$. From these two figures, we have observed that with the increasing delay, amplitude of oscillation increases. But as time progresses, the system
moves to its infected equilibrium. In the bottom panel of Fig. 5.7c and d, we keep

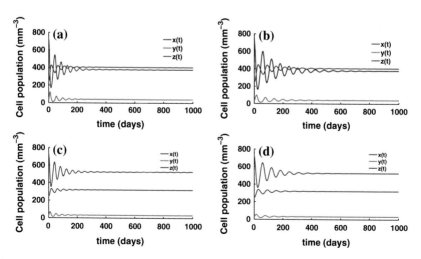

Fig. 5.8 Time series solution of the delayed system with different k and τ. Other parameters remain
unchanged as in Table 5.3

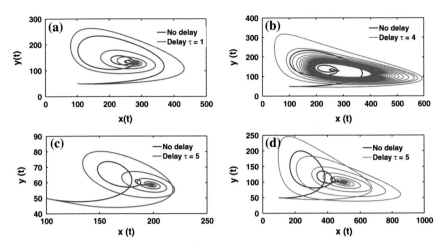

Fig. 5.9 Phase portraits for the nondelayed and delayed systems. Various model parameters are as in Table 5.3

$\beta_2 = 0.001$. From these two figures, we conclude that delay does not make any significant impact if the killing rate of infected T cells by CTLs (k) is large. From these simulations, we may conclude that the effect of delay is more important, if the CTL response is negligible in the system. Figure 5.9 shows phase plots of the delay-induced system in the model given by Eq. (5.43). In Fig. 5.9a, $\tau = 1$, $\beta_2 = 0.0005$; in Fig. 5.9b, $\tau = 5$, $\beta_2 = 0.0005$; in Fig. 5.9c, $\tau = 1$, $k = 0.001$; and in Fig. 5.9d $\tau = 5$, $\beta_2 = 0.001$. From Fig. 5.9a and b, we observe that if the delay exceeds its threshold value $\tau = 3.7709$, then the system exhibits a periodic oscillation. But if $k\beta_2$ increases, then the endemic system attains its stability (Fig. 5.9c and d). Thus the conclusions of our numerical analysis can be briefly summarized by saying that increasing the delay in the system causes increasing oscillations and instability, when CTL response against the infected cells is suppressed.

5.4 Effect of Delay in Presence of Positive Feedback Control

In this section, we assume that the production of CTLs by infected cells obeys the positive feedback control strategy and since the process is not instantaneous, there exists a finite time lag. So we incorporate the delay term in the positive feedback control function of the model (3.22).

Introducing delay term, the model equation becomes,

$$\dot{x} = \lambda + rx\left(1 - \frac{x}{c}\right) - \beta_1 xy - dx,$$
$$\dot{y} = \beta_1 xy - ay - \rho yz,$$
$$\dot{z} = \frac{ky^n(t - \tau)}{1 + ky^n(t - \tau)} - bz. \qquad (5.60)$$

For mathematical simplicity, we take $n = 1$ under the initial values $x(0) \geq 0, y(0) \geq 0, z(0) \geq 0$, where $f\{y(t - \tau)\} = \frac{ky(t-\tau)}{1+ky(t-\tau)}$. All the parameters are the same as in system (3.22) except that the positive constant τ represents the length of the delay in days.

Again we find an uninfected steady state $E_1 = (x_1, 0, 0)$ and an infected steady state $E^* = (x^*, y^*, z^*)$, where x_1, x^*, y^*, z^* are the same as in subsection (3.3), given by Eq. (3.24). Since the uninfected steady state E_1 is unstable, when $\tau = 0$ and $R_0 > 1$. Incorporation of a delay will not change the instability. Thus E_1 is unstable if $R_0 > 1$, which is also the feasibility condition for the infected steady state E^*.

To study the stability of the steady states E^*, let us define $\bar{x}(t) = x(t) - x^*$, $\bar{y}(t) = y(t) - y^*$, $\bar{z}(t) = z(t) - z^*$. Then the linearized system of (5.60) at E^* is given by,

$$\dot{\bar{x}} = \left(r - \frac{2rx^*}{c} - \beta_1 y^* - d\right)\bar{x} - \beta_1 x^* \bar{y},$$
$$\dot{\bar{y}} = \beta_1 y^* \bar{x} + \bar{y}\left(\beta_1 x^* - a - \rho z^*\right) - \rho y^* \bar{z},$$
$$\dot{\bar{z}} = f'\{y^*\}\bar{y}(t - \tau) - b\bar{z}. \qquad (5.61)$$

The characteristic equation of system (5.61)

$$\Delta(\xi) = |\ \xi I - \theta_1 - e^{-\xi\tau}\theta_2\ | = 0,$$

can be written as,

$$\xi^3 + \xi^2(b - M - N) + \xi(-bN - Mb + MN + \beta_1^2 x^* y^*)$$
$$+ (MNb + b\beta_1^2 x^* y^*) + [(\xi - M)(\rho y^* f'(y^*))]e^{-\xi\tau} = 0. \qquad (5.62)$$

Again (5.62) can be expressed as,

$$\psi(\xi, \tau) = \xi^3 + A_1 \xi^2 + A_2 \xi + A_3 e^{-\xi\tau} + (A_4 \xi)e^{-\xi\tau} + A_5 = 0. \qquad (5.63)$$

Here,

$$A_1 = b - M - N,$$
$$A_2 = -bN - Mb + MN + \beta_1^2 x^* y^*,$$
$$A_3 = -\rho y^* f'(y^*) M,$$
$$A_4 = \rho y^* f'(y^*),$$
$$A_5 = MNb + b\beta_1^2 x^* y^*.$$

For $\tau = 0$,

$$\psi(\xi, 0) = \xi^3 + A_1 \xi^2 + (A_2 + A_4)\xi + (A_3 + A_5) = 0. \tag{5.64}$$

According to the Routh–Hurwitz criterion, the nondelayed system (5.64) is locally asymptotically stable following the conditions obtained in Theorem 3.2.

Now if $\tau > 0$, then the characteristic equation (5.63) is a transcendental equation and it has infinitely many eigenvalues. So the Routh–Hurwitz criterion cannot be used for this equation. Here we try to find out the conditions of stability for the infected steady state with a finite delay.

Let $\xi(\tau) = u(\tau) + iw(\tau)$; $(w > 0)$. Since the equilibrium E^* is stable, it follows that when $\tau = 0, u(0) < 0$. Now, if $\tau > 0, u(\tau) < 0$, then E^* is still stable. If $u(\tau_0) = 0$ for certain $\tau_0 > 0$, then the steady state E^* loses its stability and it becomes unstable, when $u(\tau)$ becomes positive. If $w(\tau_0)$ does not exist, then the characteristic equation (5.63) does not have purely imaginary root for all delay, then the steady state E^* is always stable. We shall show that the above statement is true for the characteristic equation (5.63).

Suppose $\xi = iw(\tau)$ is a root of the equation (5.63) if and only if,

$$-iw^3 - A_1 w^2 + iA_2 w + (A_4 iw + A_3)(\cos w\tau - i \sin w\tau) + A_5 = 0. \tag{5.65}$$

Separating real and imaginary parts, we obtain the following transcendental equations:

$$w^3 - A_2 w = A_4 w \cos w\tau - A_3 \sin w\tau. \tag{5.66}$$

$$A_1 w^2 - A_5 = A_3 \cos w\tau + A_4 w \sin w\tau. \tag{5.67}$$

Squaring and adding the above two equations we get,

$$w^6 + (A_1^2 - 2A_2)w^4 + (A_2^2 - 2A_1 A_5 - A_4^2)w^2 + (A_5^2 - A_3^2) = 0. \tag{5.68}$$

Let $v = w^2$, then Eq. (5.68) becomes,

$$F(v) = v^3 + S_1 v^2 + S_2 v + S_3 = 0, \tag{5.69}$$

where

$$S_1 = A_1^2 - 2A_2, \quad S_2 = A_2^2 - 2A_1A_5 - A_4^2 \text{ and } S_3 = A_5^2 - A_3^2.$$

If $S_1 = A_1^2 - 2A_2 \geq 0$, $S_2 = A_2^2 - 2A_1A_5 - A_4^2 > 0$, and $S_3 = A_5^2 - A_3^2 \geq 0$, then Eq. (5.69) has no positive real root. Now,

$$\frac{dF(v)}{dv} = 3v^2 + 2S_1v + S_2 = 0. \tag{5.70}$$

Then the roots of the equation (5.70) are,

$$v_1 = \frac{-S_1 + \sqrt{S_1^2 - 3S_2}}{3} \quad \text{and} \quad v_2 = \frac{-S_1 - \sqrt{S_1^2 - 3S_2}}{3} \tag{5.71}$$

Since $S_2 > 0$, then $\sqrt{S_1^2 - 3S_2} < S_1$. Hence neither v_1 nor v_2 is positive. Thus Eq. (5.70) does not have any positive root. Since $F(0) = S_3 \geq 0$, then Eq. (5.69) has no positive root.

Remark 5.4.1 If $S_1 \geq 0$, $S_3 \geq 0$, and $S_2 > 0$, we can claim that there exists no w such that iw is the eigenvalue of the characteristic equation (5.63). Therefore, the real parts of all the eigenvalues of (5.63) are negative for all $\tau \geq 0$.

Proposition 5.4.1 *If the system satisfies*

(i) $A_1 > 0$, $A_3 + A_5 > 0$, and $A_1(A_2 + A_4) - (A_3 + A_5) > 0$,
(ii) $S_1 \geq 0$, $S_3 \geq 0$, and $S_2 > 0$ hold,

then the infected steady state E^* is asymptotically stable for all $\tau \geq 0$.

For Proposition 5.4.1, we can say that if the parameters in Table 5.1, satisfy the conditions (i) and (ii), then the infected steady state of the delay model is asymptotically stable for all delay values. Hence we can say for the above conditions, the system is stable and it is independent of the delay.

Remark 5.4.2 If $S_3 < 0$, then there exists positive root v_0 such that the characteristic equation has conjugate pair of purely imaginary roots $\pm iw_0$. For Eq. (5.69), $F(0) < 0$ and $\lim_{v \to \infty} F(v) = \infty$. Thus Eq. (5.69) has at least one positive root denoted by v_0.
Now, if $S_2 < 0$, then $\sqrt{S_1^2 - 3S_2} > S_1$ and so $v_1 = \frac{-S_1 + \sqrt{S_1^2 - 3S_2}}{3} > 0$, which implies that Eq. (5.68) has a pair of purely imaginary roots $\pm iw_0$.
Now for $\tau = \tau_0$, $u(\tau_0) = 0$, and $w(\tau_0) = w_0$ From Eqs. (5.66) and (5.67) we get,

$$\tau_n = \frac{1}{w_0} \arccos\left[\frac{(A_4w_0^4) - (A_2A_4 - A_1A_3)w_0^2 - A_3A_5}{A_4^2w_0^2 + A_3^2}\right] + \frac{2n\pi}{w_0}, \quad n = 0, 1, 2, \dots \tag{5.72}$$

where $w_0^2 = \frac{-S_1 + \sqrt{S_1^2 - 3S_2}}{3}$. Note that if the following transversality condition $\frac{d}{d\tau} Re\xi(\tau) \|_{\tau=\tau_0} = \frac{d}{d\tau}(u(\tau)) \|_{\tau=\tau_0} > 0$ holds, then the real part of $\xi(\tau)$ becomes positive, when $\tau > \tau_0$ and the steady state becomes unstable.

Proposition 5.4.2 *Suppose that*

(i) $A_1 > 0$, $A_3 + A_5 > 0$ and $A_1(A_2 + A_4) - (A_3 + A_5) > 0$. If either
(ii) $S_3 < 0$ or
(iii) $S_3 \geq 0$ and $S_2 < 0$,

then E^* is asymptotically stable, when $\tau < \tau_0$ and unstable, when $\tau > \tau_0$, where

$$\tau_0 = \frac{1}{w_0} \arccos \left[\frac{(A_4 w_0{}^4) - (A_2 A_4 - A_1 A_3) w_0{}^2 - A_3 A_5}{A_4^2 w_0^2 + A_3^2} \right], \quad n = 0. \qquad (5.73)$$

Thus, when $\tau = \tau_0$, Hopf bifurcation occurs, i.e., a family of periodic solution bifurcates for E^* as τ passes through the critical value τ_0.

5.4.1 Numerical Analysis of the Delayed System

In Fig. 5.10, we see that when $\tau = 1$, the system oscillates initially and as time increases, the system moves towards its stable region. If we consider $\tau = 3$, the amplitude of oscillation increases but after certain time, the system moves to its stable region. It has been also observed that as τ increases, the system takes more time to move towards its stable region. Figure 5.11a–c is a phase plane, where the model variables are plotted against τ. In this figure, we see that when τ increases the infectible cell population increases, whereas the infected cell population decreases rapidly. As n increases, the infectible cell population increases and the virus-producing CD4$^+$T population decreases in presence of delay.

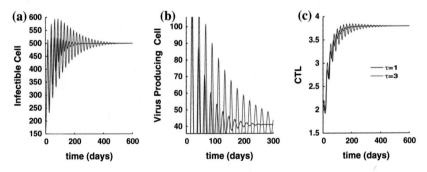

Fig. 5.10 Solution trajectories of the system for different values of τ, when $n = 1$ and keeping all other parameters are same as in Table 5.4

Table 5.4 Parameters used in the model (5.60)

Parameter	Definition	Default value assigned (day^{-1})
λ	Constant rate of production of infectible CD4$^+$T cells	10.0 mm^3
r	Proliferation rate constant of infectible CD4$^+$T cells	0.108 mm^3
c	Maximum proliferation of infectible CD4$^+$T cells	1500 mm^3
d	Death rate of infectible CD4$^+$T cells	0.01 mm^3
β	Rate of contact between x and y	0.002 mm^3
a	Death rate of virus-producing cells	0.24 mm^3
ρ	Killing rate by virus-producing cells	0.2
k	Equilibrium constant	0.001–0.005
b	Death rate of CTL responses	0.02

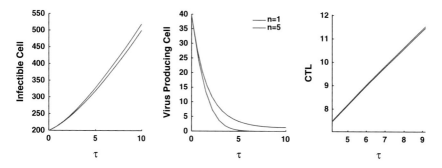

Fig. 5.11 Phase plane portrait of the solution trajectories as a function of delay

5.5 Effect of Delay during Combination of Drug Therapy

In this section, we have proposed and analyzed the mathematical model (2.11), incorporating delay in activation of uninfected CD4$^+$T cell population through $IL-2$ therapy. It should be mentioned here that delay differential equations demonstrate in a complex dynamics rather than ordinary differential equations in view of the fact that a time lag could cause a stable equilibrium to become unstable and hence the population may be fluctuated. For better understanding and also for realistic emulation of the delay-induced system, we have thus introduced a time delay in the production of CTLs in our model (2.11). We have also initiated another discrete time delay due to the account of the time lag in CTL_p activation. Thus we have the following delay differential equation model in the form of:

$$\dot{x}(t) = \lambda - \beta_1(1 - \eta_1 u_1)x(t)y(t) - dx(t) + \gamma x(t - \tau_1),$$
$$\dot{y}(t) = \beta_1(1 - \eta_1 u_1)x(t)y(t) - ay(t) - py(t)z(t),$$
$$\dot{w}(t) = cx(t)y(t)w(t) - cq_1 y(t)w(t) - bw(t) + \gamma_1 w(t - \tau_2),$$
$$\dot{z}(t) = cq_2 y(t)w(t) - hz(t), \tag{5.74}$$

with initial conditions $x(\theta) = x_0 > 0, y(\theta) = 0, w(\theta) = 0, z(\theta) = 0$ for $\theta \in [-max\{\tau_1, \tau_2\}, 0]$.

5.5.1 Stability Analysis of the Delay-Induced System

We further pay our attention to investigate the local asymptomatic stability of the infected steady state E^* for the delay-induced system (Eq. 5.74). Now linearizing the system (5.74) about E^* we get,

The characteristic equation of system (5.74) is given by,

$$\Delta(\xi) = | \xi I - F - e^{-\xi\tau_1}G - e^{-\xi\tau_2}H | = 0. \tag{5.75}$$

This equation can be written as,

$$\psi(\xi, \tau_1, \tau_2) = \xi^4 + A_1\xi^3 + A_2\xi^2 + A_3\xi + A_4 + e^{-\xi\tau_1}[B_1\xi^3 + B_2\xi^2 + B_3\xi + B_4]$$
$$+ e^{-\xi\tau_2}[C_1\xi^3 + C_2\xi^2 + C_3\xi + C_4] + e^{-\xi(\tau_1+\tau_2)}[D_1\xi^2 + D_2\xi + D_3] = 0. \tag{5.76}$$

The coefficients are given below,

$$m_{11} = -d - \beta_1(1 - \eta_1 u_1)y^*, \quad m_{12} = -\beta_1(1 - \eta_1 u_1)x^*, \quad m_{21} = \beta_1(1 - \eta_1 u_1)y^*,$$
$$m_{22} = \beta_1(1 - \eta_1 u_1)x^* - a - pz^*, \quad m_{24} = py^*, \quad m_{31} = cy^*w^*, \quad m_{32} = cx^*w^* - cq_1 w^*,$$
$$m_{33} = cx^*y^* - cq_1 y^* - b, \quad m_{42} = cq_2 w^*, m_{43} = cq_2 y^*, m_{44} = -h,$$

where

$A_1 = m_{44} - m_{33} - m_{22} - m_{11},$

$A_2 = m_{33}m_{44} + m_{11}m_{44} + m_{22}m_{44} + m_{22}m_{33} + m_{11}m_{33} + m_{22}m_{11} - m_{12}m_{21} - m_{24}m_{42},$

$A_3 = -m_{11}m_{33}m_{44} - m_{22}m_{33}m_{44} - m_{11}m_{22}m_{44} + m_{12}m_{21}m_{44} - m_{24}m_{32}m_{43}$

$\quad + m_{24}m_{33}m_{42} + m_{11}m_{24}m_{42} + m_{11}m_{24}m_{42} - m_{11}m_{22}m_{33} + m_{12}m_{21}m_{33},$

$A_4 = m_{11}m_{22}m_{33}m_{44} - m_{12}m_{21}m_{33}m_{44} + m_{11}m_{24}m_{32}m_{43} - m_{12}m_{24}m_{31}m_{43}$

$\quad - m_{11}m_{24}m_{33}m_{42},$

$B_1 = -\gamma, \quad B_2 = \gamma(m_{44} + m_{33}m_{22}), \quad B_3 = \gamma(-m_{33}m_{44} - m_{22}m_{44} - m_{22}m_{33} + m_{24}m_{42}),$

$B_4 = \gamma(m_{22}m_{33}m_{44} + m_{24}m_{32}m_{43} - m_{24}m_{33}m_{42}),$

$C_1 = -\gamma_1, \quad C_2 = \gamma_1(m_{44} + m_{22}m_{11}), \quad C_3 = \gamma_1(-m_{22}m_{44} - m_{11}m_{44} - m_{22}m_{11} + m_{24}m_{42}),$

$C_4 = \gamma_1(m_{11}m_{22}m_{44} - m_{12}m_{21}m_{44} - m_{11}m_{24}m_{42}),$

$D_1 = \gamma\gamma_1, \quad D_2 = \gamma\gamma_1(-m_{22}m_{44})$ and $D_3 = \gamma\gamma_1(m_{22}m_{44}).$

The characteristic equation (5.76) is a transcendental equation in ξ. It is known that E^* is locally asymptotically stable, if all the roots of the corresponding characteristic equation have negative real parts and unstable if purely imaginary roots appear. As we know that the transcendental equation has infinitely many complex roots, so in presence of τ_1 and τ_2, analysis of the sign of roots is very complicated. Thus, we begin our analysis by setting one delay, which is equal to zero and then deduce the conditions for stability, when both the time delays are non zero.

Case I: When $\tau_1 = \tau_2 = 0$:
In absence of both the delays, the characteristics equation (5.76) becomes,

$$\xi^4 + \xi^3(A_1 + B_1 + C_1) + \xi^2(A_2 + B_2 + C_2 + D_1) + \xi(A_3 + B_3 + C_3 + D_2)$$
$$+ (A_4 + B_4 + C_4 + D_3) = 0.$$

$$(5.77)$$

Employing Routh–Hurwitz criteria for sign of roots, we have the same results as in nondelayed system analysis (Figs. 5.12 and 5.13).

Case II: When $\tau_1 > 0, \tau_2 = 0$:
In this case, we consider no delay in CTL precursor immune response, i.e., $\tau_2 = 0$, then the characteristic equation becomes,

$$\xi^4 + \xi^3(A_1 + C_1) + \xi^2(A_2 + C_2) + \xi(A_3 + C_3) + (A_4 + C_4)$$
$$+ e^{-\xi\tau_1}[B_1\xi^3 + \xi^2(B_2 + D_1) + \xi(B_3 + D_2) + (B_4 + D_3)] = 0. \qquad (5.78)$$

For $\tau_1 > 0$, (5.78) has infinitely many roots. Using Rouche's Theorem and continuity of τ_1, the transcendental equation has roots with positive real parts if and only if it has purely imaginary roots. Let $i\theta$ be a root of equation (5.78) and hence we get,

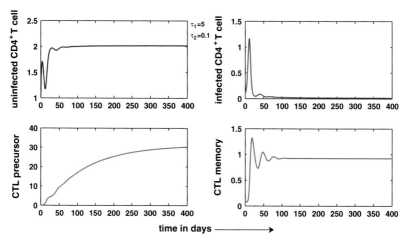

Fig. 5.12 Solution trajectory of the delayed system. Here $\tau_1 = 5$, $\tau_2 = 0.1$, and all other parameter values are same as Table 5.5

Table 5.5 Parameters used in the model (5.74)

Parameter	Definition	Value assigned
λ	Constant rate of production of CD4$^+$T cells	10.0 cells/day
d	Death rate of uninfected CD4$^+$T cells	0.01 cells/day
β	Rate of infection	0.001 cells/day
a	Death rate of infected cells	0.24 cells/day
p	Clearance rate of infected cells by CTL_{es}	0.002/day
c	Rate of proliferation of CTL_{ps}	0.6/day
b	Decay rate of CTL_{ps}	0.01/day
h	Decay rate of CTL_{es}	0.02/day
γ	Activation rate of uninfected CD4$^+$T cells by IL-2	0.5/day
γ_1	Activation rate of CTL_{ps} by IL-2	0.1/day
q_1	Multiplication capacity of differentiated precursor CTLs	0.5
q_2	Multiplication capacity of proliferated effector CTLs	0.3

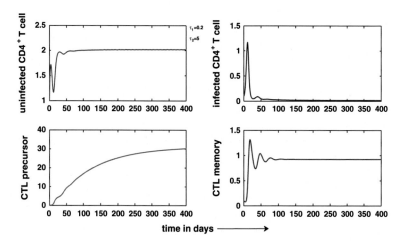

Fig. 5.13 Solution trajectory of the delayed system. Here $\tau_1 = 0.2$, $\tau_2 = 5$ and all other parameter values are same as Table 5.5

$$\theta^4 - \theta^2(A_2 + C_2) + (A_4 + C_4) = \cos\theta\tau_1[\theta^2(B_2 + D_1) - (B_4 + D_3)]$$
$$+ \sin\theta\tau_1[\theta^3 D_1 - \theta(B_3 + D_2)] \quad \text{and}$$
$$\theta(A_3 + C_3) - \theta^3(A_1 + C_1) = \cos\theta\tau_1[\theta^3 D_1 - \theta(B_3 + D_2)]$$
$$- \sin\theta\tau_1[\theta^2(B_2 + D_1) - (B_4 + D_3)].$$

$$(5.79)$$

Squaring and adding above two equations we have,

$$
\begin{aligned}
&\theta^8 + \theta^6[(A_1 + C_1)^2 - 2(A_2 + C_2) - D_1^2] \\
&+ \theta^4[(A_2 + C_2)^2 + 2(A_4 + C_4) - 2(A_3 + C_3)(A_1 + C_1) - (B_2 + D_1)^2 \\
&+ 2D_1(B_3 + D_2)] + \theta^2[(A_3 + C_3)^2 \\
&- 2(A_2 + C_2)(A_4 + C_4) + 2(B_2 + D_1)(B_4 + D_3) - (B_3 + D_2)^2] \\
&+ [(A_4 + C_4)^2 - (B_4 + D_3)^2] = 0.
\end{aligned}
$$
$$(5.80)$$

Simplifying and substituting $\theta^2 = l$ in Eq. (5.80), we get the following equation:

$$l^4 + \alpha_1 l^3 + \alpha_2 l^2 + \alpha_3 l + \alpha_4 = 0, \tag{5.81}$$

where

$$
\begin{aligned}
\alpha_1 &= (A_1 + C_1)^2 - 2(A_2 + C_2) - D_1^2, \\
\alpha_2 &= (A_2 + C_2)^2 + 2(A_4 + C_4) - 2(A_3 + C_3)(A_1 + C_1) - (B_2 + D_1)^2 + 2D_1(B_3 + D_2), \\
\alpha_3 &= (A_3 + C_3)^2 - 2(A_2 + C_2)(A_4 + C_4) + 2(B_2 + D_1)(B_4 + D_3) - (B_3 + D_2)^2, \\
\alpha_4 &= (A_4 + C_4)^2 - (B_4 + D_3)^2.
\end{aligned}
$$

It may be noted that Eq. (5.81) will have negative real part if and only if Routh–Hurwitz criterion is satisfied and hence Eq. (5.78) will have no purely imaginary root. From the above analysis, we have the following proposition:

Proposition 5.5.1 *In the delay-induced system (5.74), the infected steady state* E^* *will be locally asymptotically stable for all* $\tau_1 > 0$, *if the following conditions are satisfied:*

$$\alpha_1 > 0, \ \alpha_4 > 0, \ \psi = \alpha_1\alpha_2 - \alpha_3 > 0 \text{ and } \psi\alpha_3 - \alpha_1^2\alpha_4 > 0.$$

Case III: When $\tau_1 = 0, \tau_2 > 0$:
In absence of τ_1, the characteristic equation (5.76) have the following form:

$$
\begin{aligned}
&\xi^4 + \xi^3(A_1 + B_1) + \xi^2(A_2 + B_2) + \xi(A_3 + B_3) + (A_4 + B_4) \\
&+ e^{-\xi\tau_2}[C_1\xi^3 + \xi^2(C_2 + D_1) + \xi(C_3 + D_2) + (C_4 + D_3)] = 0. \quad (5.82)
\end{aligned}
$$

Similarly as in Case II, we substitute $\xi = i\theta$ and we get,

$$
\begin{aligned}
\theta^4 - \theta^2(A_2 + B_2) + (A_4 + B_4) &= \cos\theta\tau_2[\theta^2(C_2 + D_1) - (C_4 + D_3)] \\
&+ \sin\theta\tau_2[\theta^3 C_1 - \theta(C_3 + D_2)] \quad \text{and} \quad (5.83)
\end{aligned}
$$

$$\theta(A_3 + B_3) - \theta^3(A_1 + B_1) = \cos\theta\tau_2[\theta^3 C_1 - \theta(C_3 + D_2)]$$
$$- \sin\theta\tau_1[\theta^2(C_2 + D_1) - (C_4 + D_3)]. \qquad (5.84)$$

Squaring, adding, and then substituting $\theta^2 = s$ we have,

$$s^4 + \delta_1 s^3 + \delta_2 s^2 + \delta_3 s + \delta_4 = 0, \qquad (5.85)$$

where

$$\delta_1 = (A_1 + B_1)^2 - 2(A_2 + B_2) - C_1^2,$$
$$\delta_2 = (A_2 + B_2)^2 + 2(A_4 + B_4) - 2(A_3 + B_3)(A_1 + B_1) - (C_2 + D_1)^2 + 2C_1(C_3 + D_2),$$
$$\delta_3 = (A_3 + B_3)^2 - 2(A_2 + B_2)(A_4 + B_4) + 2(C_2 + D_1)(C_4 + D_3) - (C_3 + D_2)^2,$$
$$\delta_4 = (A_4 + B_4)^2 - (C_4 + D_3)^2.$$

From the above analysis, we have the following proposition:

Proposition 5.5.2 *In the delay-induced system (5.74), the infected steady state E^* will be locally asymptotically stable for all $\tau_2 > 0$, if the following conditions are satisfied:*

$$\delta_1 > 0, \ \delta_4 > 0, \ \varphi = \delta_1\delta_2 - \delta_3 > 0 \text{ and } \varphi\delta_3 - \delta_1^2\delta_4 > 0.$$

If $\delta_4 < 0$, then we have the following proposition:

Proposition 5.5.3 *Equation (5.85) admits at least one positive root, if $\delta_4 < 0$ is satisfied.*

If θ_0 be a positive root of (5.85), then Eq. (5.82) will have a purely imaginary root $\pm i\theta_0$ corresponding to τ_2. Now we evaluate the critical value of τ_2 for which the delay induced system (5.74) remains stable. From Eqs. (5.83) and (5.84), we obtain,

$$\tau_2^* = \frac{a\cos\phi(\theta_0)}{\theta_0}, \qquad (5.86)$$

where

$$\phi(\theta_0) = [\{\theta^2(C_2 + D_1) - (C_4 + D_3)\}\{\theta^4 - \theta^2(A_2 + B_2) + (A_4 + B_4)\}$$
$$+ \{\theta^3 C_1 - \theta(C_3 + D_2)\}\{\theta(A_3 + B_3) - \theta^3(A_1 + B_1)\}]$$
$$\div [\{\theta^2(C_2 + D_1) - (C_4 + D_3)\}^2 + \{\theta^3 C_1 - \theta(C_3 + D_2)\}^2]. \quad (5.87)$$

From the above analysis, we construct the following theorem:

Theorem 5.5.4 *If $\delta_4 < 0$ is satisfied, then the steady state E^* is locally asymptotically stable for $\tau_2 < \tau_2^*$ and becomes unstable for $\tau_2 > \tau_2^*$. When $\tau_2 = \tau_2^*$, a Hopf bifurcation occurs.*

Case IV: When $\tau_1 > 0$, $\tau_2 > 0$:

In this case, we have studied the stability of the steady state E^* in the presence of both the delays. If all the roots of equation (5.82) have negative real parts for $\tau > 0$, i.e., when the system is locally asymptotically stable, then there exists a τ_1^* depending upon τ_2 such that all roots of equation (5.76) have negative real parts, whenever $\tau_1 < \tau_1^*$. Considering all the cases, we have the following theorem:

Theorem 5.5.5 *If $\delta_4 < 0$ holds, then for $\tau_2 < \tau_2^*$, there exists a τ_1^* depending upon τ_2, the steady state E^* is locally asymptotically stable for $\tau_1 < \tau_1^*$ and $\tau_2 < \tau_2^*$.*

Special Remarks of the Delay Induced System in view of Numerical Analysis

Though it is eventually true from our analytical results that for a longer value of delay. The system becomes unstable (Theorem 1 and Theorem 2), however it is essential to mention here that in our delay-induced system, numerical analysis reveals that when $\tau_1 = 0$ and $\tau_2 = 1$. it reflects from solution trajectory, there is no such oscillation in figure, which may conclude that delay does affect the stability of the system.

Further we change the values of both the delays τ_1 and τ_2 and varied them from 1 to 20. We also observe that no significant change does arise in each case for which we can express that delay affects the stability of the system. Thus, in a nutshell we can say that incorporation of time delay into the existing model to account for time lag in activation of CTL_p does not exhibit any biologically significant interpretation.

5.6 Delay-Induced System in Presence of Saturation Effect

The proliferation of CTL in our immune system is partially dependent on the infected population cell density and since the process is simultaneous but not instantaneous, we have incorporated a realistic time lag in the production term of CTL of our basic mathematical model (2.12) of the HIV-1 infection. Thus introducing delay term in the model equation which becomes,

$$\dot{x} = \lambda - \beta_1 xy - cx,$$
$$\dot{y} = \beta_1 xy - by - \frac{\alpha yz}{a+y},$$
$$\dot{z} = \frac{uy(t-\tau)z(t-\tau)}{a+y(t-\tau)} - dz. \tag{5.88}$$

The linearized form of the system (5.88) about (x^*, y^*, z^*) becomes,

$$\dot{P} = (-\beta_1 y^* - c)P(t) - \beta_1 x^* Q(t),$$
$$\dot{Q} = \beta_1 y^* P(t) + (\beta_1 x^* - b - \frac{\alpha az^*}{(a+y^*)^2})Q(t) + \frac{\alpha y^*}{a+y^*}R(t),$$
$$\dot{R} = \frac{auz^*}{(a+y)^2}Q(t-\tau) + \frac{uy^*}{a+y^*}R(t-\tau) - dR(t). \tag{5.89}$$

The characteristic equation of system (5.89) is given by,

$$\xi^3 + (A_{d1})\xi^2 + (A_{d2})\xi + (A_{d5}\xi^2 + A_{d4}\xi + A_{d3})e^{-\xi\tau} + A_{d6} = 0, \qquad (5.90)$$

which can be written as,

$$\psi(\xi,\tau) = \xi^3 + A_{d1}\xi^2 + A_{d2}\xi + A_{d3}e^{-\xi\tau} + (A_{d4}\xi)e^{-\xi\tau} + (A_{d5}\xi^2)e^{-\xi\tau} + A_{d6} = 0, \qquad (5.91)$$

where

$A_{d1} = m_{33} - m_{22} - m_{11}, A_{d2} = -m_{22}m_{33} - m_{11}m_{33} + m_{11}m_{22} - m_{12}m_{21}, A_{d3} = -m_{11}m_{22}m_{33} + m_{11}m_{23}m_{32} + m_{12}m_{21}m_{33}, A_{d4} = m_{22}m_{33} - m_{23}m_{32} + m_{11}m_{33}, A_{d5} = -m_{33}, \quad A_{d6} = m_{11}m_{22}m_{33} - m_{12}m_{21}m_{33},$

where

$$m_{11} = \frac{-\beta_1 da}{u-d} - c, \quad m_{12} = -\frac{\beta_1\lambda(u-d)}{ad\beta_1 + c(u-d)}, \quad m_{21} = \frac{\beta_1 da}{u-d},$$

$$m_{22} = \frac{\alpha dz^*(u-d)}{u^2 a}, \quad m_{23} = -\frac{\alpha d}{u}, \quad m_{32} = \frac{z^*(u-d)^2}{ua}, \quad m_{33} = d.$$

For $\tau = 0$,

$$\psi(\xi,0) = \xi^3 + (A_{d1} + A_{d5})\xi^2 + (A_{d2} + A_{d4})\xi + (A_{d3} + A_{d6}) = 0, \qquad (5.92)$$

which is actually same as the equation $\xi^3 + A_1\xi^2 + A_2\xi + A_3 = 0$ of Proposition 2.5.3, since $A_1 = A_{d1} + A_{d5}, A_2 = A_{d2} + A_{d4}$, and $A_3 = A_{d6} + A_{d3}$.

According to the Routh–Hurwitz criterion, the nondelayed system (5.92) has all eigenvalues with negative real parts if and only if $A_1 > 0, A_3 > 0$, and $A_1A_2 - A_3 > 0$, which is equivalent with the condition obtained in Proposition 2.5.3. Now if $\tau > 0$, then the characteristic equation (5.91) is a transcendental equation and it has infinitely many eigenvalues. So the Routh–Hurwitz criterion cannot be used for this equation. Here we try to find out the conditions of stability for the infected steady state with a finite delay.

Let $\xi(\tau) = \delta(\tau) + i\omega(\tau) \quad (\omega > 0)$.

Since the equilibrium E^* is stable, it follows that $\tau = 0$ and $\delta(0) < 0$ holds. Now if $\tau > 0$ and $\delta(\tau) < 0$, then E^* is still stable. If $\delta(\tau_0) = 0$ for certain $\tau_0 > 0$, then the steady state E^* loses its stability and it becomes unstable, when $\delta(\tau)$ becomes positive. If $\omega(\tau_0)$ does not exist, then the characteristic equation (5.91) does not have purely imaginary root for all delay and the steady state E^* is always stable. We shall show that the above statement is true for the characteristic equation (5.91). Suppose $\xi = i\omega(\tau)$ is a root of the equation (5.90) if and only if,

$$-i\omega^3 - A_{d1}\omega^2 + iA_{d2}\omega + (-A_{d5}\omega^2 + i\omega A_{d4} + A_{d3})(\cos\omega\tau - i\sin\omega\tau) + A_{d6} = 0. \qquad (5.93)$$

Separating real and imaginary parts, we obtain the following transcendental equations:

$$-\omega^3 + A_{d2}\omega = A_{d4}\omega \cos \omega\tau + (A_{d5}\omega^2 - A_{d3}) \sin \omega\tau, \tag{5.94}$$

$$-A_{d1}\omega^2 + A_{d6} = -(A_{d5}\omega^2 - A_{d3}) \cos \omega\tau + A_{d4}\omega \sin \omega\tau . \tag{5.95}$$

Squaring and adding the above two equations we get,

$$\omega^6 + (A_{d1}^2 - 2A_{d2} - A_{d5}^2)\omega^4 + (A_{d2}^2 - 2A_{d1}A_{d6} + 2A_{d3}A_{d5} - A_{d4}^2)\omega^2 + (A_{d6}^2 - A_{d3}^2) = 0. \tag{5.96}$$

Let $v = w^2$, then Eq. (5.96) becomes,

$$F(v) = v^3 + S_1v^2 + S_2v + S_3 = 0, \tag{5.97}$$

where

$$S_1 = A_{d1}^2 - 2A_{d2} - A_{d5}^2, \quad S_2 = A_{d2}^2 - 2A_{d1}A_{d6} + 2A_{d3}A_{d5} - A_{d4}^2, \quad S_3 = A_{d6}^2 - A_{d3}^2.$$

If $S_1 = A_{d1}^2 - 2A_{d2} - A_{d5}^2 \geq 0$, $S_2 = A_{d2}^2 - 2A_{d1}A_{d6} + 2A_{d3}A_{d5} - A_{d4}^2 > 0$, and $S_3 = A_{d6}^2 - A_{d3}^2 \geq 0$, then Eq. (5.97) has no positive real root.

$$\text{Now,} \quad \frac{dF(v)}{dv} = 3v^2 + 2S_1v + S_2 = 0. \tag{5.98}$$

Then the roots of the equation (5.98) are,

$$v_1 = \frac{-S_1 + \sqrt{S_1^2 - 3S_2}}{3} \quad \text{and} \quad v_2 = \frac{-S_1 - \sqrt{S_1^2 - 3S_2}}{3}. \tag{5.99}$$

Since $S_2 > 0$, then $\sqrt{S_1^2 - 3S_2} < S_1$. Hence neither v_1 nor v_2 is positive. Thus, Eq. (5.98) does not have any positive root. Since $F(0) = S_3 \geq 0$, Eq. (5.97) has no positive root.

Assumption 5.6.1 If $S_1 \geq 0$, $S_3 \geq 0$, and $S_2 > 0$, we can claim that there exists no ω such that $i\omega$ is the eigenvalue of the characteristic equation (5.91), which are negative for all delay $\tau > 0$. Therefore, the real parts of all the eigenvalues of (5.91) are negative for all delay $\tau \geq 0$.

Proposition 5.6.1 *If the system satisfies*

(i) $A_{d1} + A_{d5} > 0$, $A_{d3} + A_{d6} > 0$ *and* $(A_{d1} + A_{d5})(A_{d2} + A_{d4}) - (A_{d3} + A_{d6}) > 0$
and
(ii) $S_1 \geq 0$, $S_3 \geq 0$, *and* $S_2 > 0$ *hold, then the infected steady state* E^* *is asymptotically stable for all* $\tau \geq 0$.

For Proposition 5.6.1, we can say that if the parameters in Table 5.1 satisfy the conditions (i) and (ii), then the infected steady state of the delay model is asymptotically stable for all delay values. Hence we can say for the above two conditions, system is stable and it is thus independent of delay.

Assumption 5.6.2 If $S_3 < 0$, then there exists a positive root v_0 such that characteristic equation has conjugate pair of purely imaginary roots $\pm i\omega_0$. From Eq. (5.97), we get $F(0) < 0$ and $\lim_{v \to \infty} F(v) = \infty$ and thus Eq. (5.97) has at least one positive root, which is denoted by v_0.

Now if $S_2 < 0$, then obviously $\sqrt{S_1^2 - 3S_2} > S_1$. Thus $v_1 = \frac{-S_1 + \sqrt{S_1^2 - 3S_2}}{3} > 0$. This implies that Eq. (5.96) has a pair of purely imaginary roots which is denoted by $\pm i\omega_0$. Again $\tau = \tau_0$ shows that $\delta(\tau_0) = 0$ and $\omega(\tau_0) = \omega_0$. So from Eqs. (5.94) and (5.95) we get,

$$\tau_n = \frac{1}{\omega_0} \arccos\left[\frac{(A_{d1}A_{d5} - A_{d4})\omega_0^4 - (A_{d2}A_{d4} - A_{d1}A_{d3} - A_{d5}A_{d6})\omega_0^2 + A_{d3}A_{d6}}{A_{d5}^2\omega_0^4 + (A_{d4}^2 - 2A_{d3}A_{d5})\omega_0^2 + A_{d3}^2} \right]$$
$$+ \frac{2n\pi}{\omega_0}, n = 0, 1, 2 \ldots \ldots \tag{5.100}$$

where $w_0^2 = \frac{-S_1 + \sqrt{S_1^2 - 3S_2}}{3}$.

It should be noted here that the following transversality condition, $\frac{d}{d\tau} Re\xi(\tau)$ $\|_{\tau=\tau_0} = \frac{d}{d\tau}(\delta(\tau))\|_{\tau=\tau_0} > 0$ holds. By continuity, the real part of $\xi(\tau)$ becomes positive, when $\tau > \tau_0$ and the steady state becomes unstable. Furthermore, a Hopf bifurcation occurs, when τ passes through the critical value τ_0.

Proposition 5.6.2 *Suppose that*
(i) $A_{d1} + A_{d5} > 0$, $A_{d3} + A_{d6} > 0$ *and* $(A_{d1} + A_{d5})(A_{d2} + A_{d4}) - (A_{d3} + A_{d6}) > 0$. *If either (ii)* $S_3 < 0$ *or (iii)* $S_3 \geq 0$ *and* $S_2 < 0$, *then* E^* *is asymptotically stable, when* $\tau < \tau_0$, *and unstable when* $\tau > \tau_0$, *where*

$$\tau_0 = \frac{1}{\omega_0} \arccos\left[\frac{(A_{d1}A_{d5} - A_{d4})\omega_0^4 - (A_{d2}A_{d4} - A_{d1}A_{d3} - A_{d5}A_{d6})\omega_0^2 + A_3A_6}{A_{d5}^2\omega_0^4 + (A_{d4}^2 - 2A_{d3}A_{d5})\omega_0^2 + A_{d3}^2} \right],$$
$$n = 0. \tag{5.101}$$

Table 5.6 Parameters used in the model (5.88)

Parameter	Definition	Default value assigned
λ	Constant rate of production rate of CD4$^+$T cells	10.0 mm^{-3} Day^{-1}
c	Death rate of uninfected CD4$^+$T cells	0.007 Day^{-1}
β	Rate of contact between x and y	0.005 mm^{-3} Day^{-1}
b	Death rate of virus-producing cells	0.25 Day^{-1}
α	Killing rate of virus-producing cells	0.75 mm^{-3} Day^{-1}
a	Half saturation constant	47 mm^{-3}
u	Simulation rate of CTLs	0.27 Day^{-1}
d	Death rate of CTLs	0.1 Day^{-1}

Thus when $\tau = \tau_0$, Hopf bifurcation occurs, i.e., a family of periodic solution bifurcates for E^* as τ passes through the critical value τ_0.

Proposition 5.6.2 specifies that incorporating delay, system could exhibit Hopf bifurcation at definite value of the delay if the parameters satisfy the conditions in (ii) and (iii). However, the parameter values given in Table 5.6 do not hold neither condition (ii) nor (iii).

Here we have incorporated a time lag in our model in the third equation, since the process of generation of CTL is not instantaneous, so we have incorporated a realistic time lag in the production term of CTL of our basic mathematical model (2.12) of the HIV-1 infection. We have estimated the length of delay for which the stability of the system remains preserved. We obtain the threshold value for τ, below this critical value of τ, interior equilibrium point becomes asymptotically stable and when the value of τ exceeds this value, interior equilibrium point becomes unstable and at this critical value a Hopf bifurcation occurs.

References

1. Bonhoeffer, S., Coffin, J.M., Nowak, M.A.: Human immunodeficiency virus drug therapy and virus load. J. Virol. **71**, 3275–3278 (1997)
2. MacDonald, M.: Time Lags in Biological Models. Lecture Notes in Biomathematics, vol. 27. Springer-Verlag, New York (1978)
3. Culshaw, R.V., Ruan, S., Webb, G.: A mathematical model of cell to cell spread of HIV-1 that includes a time delay. J. Math. Biol. **46**, 425–444 (2003)
4. May, R.M.: Stability and Complexity in Model Ecosystems, 2nd edn. Princeton University Press, Princeton (1974)
5. Gopalsamy, K.: Stability and Oscillations in Delay Differential Equations of Population Dynamics. Kluwer Academic (1992)
6. Marsden, J.E., McCracken, M.: The Hopf Bifurcation and its Application. Springer, New York (1976)
7. Perelson, A.S., Neuman, A.U., Markowitz, M., Leonard, J.M., Ho, D.D.: HIV 1 dynamics in vivo: viron clearance rate, infected cell life span, and viral generation time. Science **271**, 1582–1586 (1996)
8. Yang, J., Wang, X., Zhang, F.: A differential equation model of HIV infection of CD4$^+$ T cells with delay. Disc. Dyn. Nat. Soc. Article ID 903678, 16 p (2008). doi:10.1155/2008/903678. 179
9. Li, D., Ma, W.: Asymptotic properties of a HIV-1 infection model with time delay. J. Math. Anal. Appl. **335**, 683–691 (2007)
10. Wang, K., Wang, W., Pang, H., Liu, X.: Complex dynamic behavior in a viral model with delayed immune response. Phys. D **226**, 197–208 (2007)
11. Bachar, M., Dorfmayr, A.: HIV treatment models with time delay. C. R. Biol. **327**, 983–994 (2004)
12. McCluskey, C.C.: Complete global stability for an SIR epidemic model with delay–Distributed or discrete. Nonlinear Anal.: Real World Appl. 11, 55–59 (2010)

Chapter 6
Optimal Control Theory

Abstract Exploration of the optimal control theoretic approach has been investigated. Here, our main aim is to minimize the cost as well as minimize the infected $CD4^+T$ cells and maximize the uninfected $CD4^+T$ cells. Moreover, optimal control strategy helps for successful immune reconstitution that can be achieved with increase in precursor CTL population. Mathematical modeling of viral dynamics enables the maximization of therapeutic outcome even in case of multiple therapies with specific goal of reversal of immunity impairment. From our analytical and numerical analysis, it is vibrant that implementation of optimal control approach in optimization of therapeutic regime of combination therapy of HAART and IL-2 are found to be really acceptable. This is satisfactory in terms of enhancing the life prospect of HIV-afflicted patients by improving the uninfected T cell count.

Keywords Optimal control problem · Perfect adherence · Weight constant · Objective cost functional · Hamiltonian · Pontryagin minimum principle · Adjoint variable

6.1 Optimal Control Theoretic Approach of the Implicit Model

In this section, our main aim is to minimize the cost as well as minimize the infected $CD4^+T$ cells and maximize the uninfected $CD4^+T$ cells. We construct the optimal control problem for the model (2.6), where the state system is given by,

$$\dot{x} = \lambda - d_1 x - \beta_1 (1 - u_1(t))xy_i + u_2(t)x,$$
$$\dot{y_i} = \beta_1 (1 - u_1(t))xy_i - d_2 y_i - \beta_2 y_i z + \delta y_l,$$
$$\dot{y_l} = v y_i - d_3 y_l - \delta y_l,$$
$$\dot{z} = s y_i - d_4 z + u_2(t)z, \tag{6.1}$$

© Springer Science+Business Media Singapore 2015
P.K. Roy, *Mathematical Models for Therapeutic Approaches
to Control HIV Disease Transmission*, Industrial and Applied Mathematics,
DOI 10.1007/978-981-287-852-6_6

and the control function is defined as,

$$J(u_1, u_2) = \int\limits_{t_i}^{t_f} [Pu_1{}^2 + Qu_2{}^2 - x^2 + y_i{}^2]dt. \tag{6.2}$$

The parameters P and Q are the weight on the benefit for the cost of treatment. These are the per unit cost of RTI and IL-2 respectively.

Here the control $u_1(t)$ represents the efficacy of the drug therapy inhibiting the reverse transcription, i.e., blocking new infection. The control $u_2(t)$ represents the efficacy of IL-2 therapy, i.e., immune boosting.

In this problem, we are seeking the optimal control pair $(u_1{}^*(t), u_2{}^*(t))$ such that

$$J(u_1{}^*, u_2{}^*) = \min\{J(u_1, u_2) : (u_1, u_2) \in U\}.$$

Here U is the control set defined by $U = U_1 \times U_2$, where $U_1 = U_2 = \{u(t) : u_1, u_2 \text{ are Lebesgue measurable, } 0 \le u_1(t) \le 1, \ 0 \le u_2(t) \le 1, \ t \in [t_i, t_f]\}$.

To determine the optimal control $u_1{}^*(t)$ and $u_2{}^*(t)$, we use the "Pontryagin's Minimum Principle" [1]. To solve the problem, we use the Hamiltonian [2, 3], which is given by,

$$\begin{aligned}
H = {} & Pu_1{}^2 + Qu_2{}^2 - x^2 + y_i{}^2 + \xi_1\{\lambda - d_1 x - \beta_1(1 - u_1(t))xy_i + u_2(t)x\} \\
& + \xi_2\{\beta_1(1 - u_1(t))xy_i - d_2 y_i - \beta_2 y_i z + \delta y_l\} + \xi_3\{vy_i - d_3 y_l - \delta y_l\} \\
& + \xi_4\{s y_i - d_4 z + u_2(t)z\} + v_{11}u_1(t) + v_{12}(1 - u_1(t)) + v_{21}u_2(t) \\
& + v_{22}(1 - u_2(t)),
\end{aligned} \tag{6.3}$$

where v_{ij} $(i = 1, 2; j = 1, 2)$ are penalty multipliers ensuring that $u_1(t)$ and $u_2(t)$ remain bounded in $[t_i, t_f]$. We also have $v_{11}u_1^*(t) = v_{12}(1 - u_1^*(t)) = v_{21}u_2^*(t) = v_{22}(1 - u_2^*(t)) = 0$ and $\xi_i(t), i = 1, 2, 3, 4$ are the adjoint variables.

Using the "Pontryagin's Minimum Principle" and for the existence condition of the adjoint variables [4], we obtain the following theorem:

Theorem 6.1.1 *If the objective cost function $J(u_1, u_2)$ over U is minimum for the optimal control $u^* = (u_1{}^*, u_2{}^*)$ corresponding to the endemic equilibrium $(x^*, y_i{}^*, y_l{}^*, z^*)$, then there exists adjoint variables $\xi_1, \xi_2, \xi_3,$ and ξ_4 satisfying the equations $\frac{d\xi_1}{dt} = -\frac{\partial H}{\partial x}, \frac{d\xi_2}{dt} = -\frac{\partial H}{\partial y_i}, \frac{d\xi_3}{dt} = -\frac{\partial H}{\partial y_l}, \frac{d\xi_4}{dt} = -\frac{\partial H}{\partial z}$ with the transversality conditions $\xi_i(t_f) = 0, i = 1, 2, 3, 4.$*

Proof Using Pontryagin's Minimum principle [4], the unconstrained optimal control variables $u_1{}^*(t)$ and $u_2{}^*(t)$ satisfy,

$$\frac{\partial H}{\partial u_1{}^*} = \frac{\partial H}{\partial u_2{}^*} = 0. \tag{6.4}$$

$$H = [Pu_1{}^2(t) - \xi_1(1 - u_1(t))\beta_1 x y_i + \xi_2(1 - u_1(t))\beta_1 x y_i + v_{11}u_1(t) + v_{12}(1 - u_1(t))]$$
$$+ [Qu_2{}^2(t) + \xi_1 u_2 x + \xi_4 u_2 z + v_{21}u_2(t) + v_{22}(1 - u_2(t))]$$
$$+ \text{other terms without } u_1(t) \text{ and } u_2(t), \tag{6.5}$$

we have obtained $\dfrac{\partial H}{\partial u_i{}^*}$ for $u_i{}^*(i = 1, 2)$ and equating to zero we get,

$$\frac{\partial H}{\partial u_1{}^*} = 2Pu_1{}^* + \beta_1 x y_i(\xi_1 - \xi_2) + v_{11} - v_{12} = 0,$$
$$\frac{\partial H}{\partial u_2{}^*} = 2Qu_2{}^* + \xi_1 x + \xi_4 z + v_{21} - v_{22} = 0.$$

Further, $u_1^*(t)$ and $u_2^*(t)$ can be written in the following form,

$$u_1{}^*(t) = \frac{\beta_1 x y_i(\xi_2 - \xi_1) + v_{12} - v_{11}}{2P},$$
$$u_2{}^*(t) = \frac{-(\xi_1 x + \xi_4 z) + v_{22} - v_{21}}{2Q}. \tag{6.6}$$

Since the standard control is bounded, we conclude for the control $u_1(t)$:

$$u_1^*(t) = \begin{cases} 0, & \frac{\beta_1 x y_i(\xi_2-\xi_1)}{2P} \leq 0; \\ \frac{\beta_1 x y_i(\xi_2-\xi_1)}{2P}, & 0 < \frac{\beta_1 x y_i(\xi_2-\xi_1)}{2P} < 1; \\ 1 & \frac{\beta_1 x y_i(\xi_2-\xi_1)}{2P} \geq 1. \end{cases}$$

Hence, the compact form of $u_1^*(t)$ is given by,

$$u_1^*(t) = \max\{\min\{1, \frac{\beta_1 x y_i(\xi_2 - \xi_1)}{2P}\}, 0\}. \tag{6.7}$$

In a similar way, we get the compact form of u_2^* as,

$$u_2^*(t) = \max\{\min\{1, \frac{-(\xi_1 x + \xi_4 z)}{2Q}\}, 0\}. \tag{6.8}$$

Using the Corollary 4.1 of Fleming and Rishel [4], the existence of the optimal control due to the convexity of the integrand J with respect to u_1 and u_2, a boundedness and the Lipschitz property of the state system with respect to state variables, the adjoint system can be written as, $\frac{d\xi_1}{dt} = -\frac{\partial H}{\partial x}$, $\frac{d\xi_2}{dt} = -\frac{\partial H}{\partial y_i}$, $\frac{d\xi_3}{dt} = -\frac{\partial H}{\partial y_l}$ and $\frac{d\xi_4}{dt} = -\frac{\partial H}{\partial z}$. Taking the partial derivative of H, we get the following adjoint equations,

$$\frac{d\xi_1}{dt} = 2x + \xi_1\{d_1 + (1 - u_1(t))\beta_1 y_i - u_2(t)\} - \xi_2(1 - u_1(t))\beta_1 y_i,$$

$$\frac{d\xi_2}{dt} = -2y_i + \xi_1(1 - u_1(t))\beta_1 x - \xi_2\{(1 - u_1(t))\beta_1 x - d_2 - \beta_2 z\} - \xi_3 v - \xi_4 s,$$

$$\frac{d\xi_3}{dt} = -\xi_2 \delta + \xi_3(d_3 + \delta),$$

$$\frac{d\xi_4}{dt} = \xi_2 \beta_2 y_i + \xi_4(d_4 - u_2(t)). \tag{6.9}$$

The optimality of the system consists the state system with the adjoint system together with the initial conditions. The transversality conditions satisfying $\xi_i(t_f) = 0$, ($i = 1, 2, 3, 4$) and $x(0) = x_0$, $y_i(0) = y_{i0}$, $y_l(0) = y_{l0}$, $z(0) = z_0$.

6.1.1 Numerical Simulation of the Implicit Model

In Table 6.1, all parameter values are default values and these values are collected from different journals [5–9]. In the presence of virus, the system is very much inconsistent and it is very difficult to choose the parameter values. Many of the parameters have not been measured and assumed from different ranges. We have considered the new CD4$^+$T cells are migrated from the thymus at a rate $\lambda = 10$ day^{-1} mm^{-3} [7]. Since T cells have its natural life span, it is assumed that its natural death rate $d_1 = 0.02$ day^{-1} mm^{-3} [6, 7]. It has also been observed that the rate of infection is $0.0025 \sim 0.5$ [10–12].

Biomedical or virological studies [13] show that the natural life of infected cells are 2–6 weeks in the absence of antigen-stimulated replication. On that outlook, we have assumed that the natural death rates of infected CD4$^+$T cells are $d_2 = 0.2$ day^{-1}mm^{-3}. In the presence of immune response (or in the absence of immune impairment effect) the stimulation rate of CTLs (s), the rate at which uninfected CD4$^+$T cells are killed by CTLs (β_2) and natural death rate of drug induced CTLs (i.e., d_4) is restricted by the constraint $\frac{s\beta_2}{d_4} \sim 0.01 - 0.05$ [9]. From this restriction, we assume the default values of s, β_2 and d_4 mentioned in Table 6.1.

We use the default values of the parameters given in Table 6.1. We have also observed that for another set of parameters with their reported range can be used and give the similar behavior.

In Fig. 6.1, we choose $u_1 = 0.5$ and $u_2 = 0$. This means, there is no IL-2 therapy and only RTI is administered. It is clearly seen that the concentration of uninfected

Table 6.1 List of parameters for system (2.6)

Parameter	d_1	d_2	d_3	d_4	β_2	δ	v	s
Default value	0.03	0.3	0.03	0.2	0.001	0.1	0.5	0.2

Units are mm^{-3}day^{-1} except β_2, s, and d_4 as because there units are day^{-1}

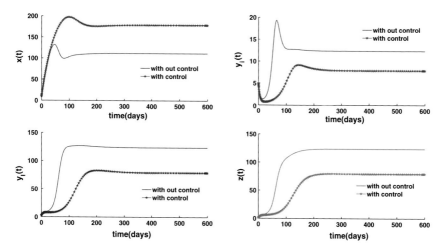

Fig. 6.1 The system behavior for without control ($u_1 = 0$ and $u_2 = 0$) and with control ($u_1 = 0.5$ and $u_2 = 0$)

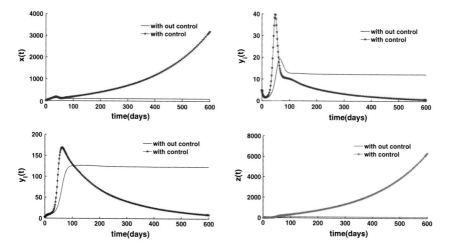

Fig. 6.2 The system behavior for without control ($u_1 = 0$ and $u_2 = 0$) and with control ($u_1 = 0$ and $u_2 = 0.025$)

CD4$^+$T cells in the presence of treatment reaches a higher steady state compared to in the absence of treatment. Furthermore, the infected, latently infected CD4$^+$T cells and CTL response in the presence of treatment reach a lower steady state compared to the absence of treatment. However, if we only bring in the IL-2 and RTIs are not been introduced (Fig. 6.2), then the uninfected CD4$^+$T cells and CTL responses concentration increase rapidly. However, the infected and latently infected CD4$^+$T cells increase within 100 days. After that, its concentration suddenly decreases and as time progresses, their concentration moves towards extinction. Infected and latently

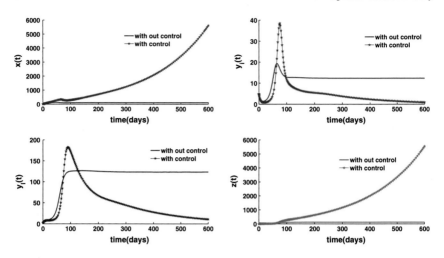

Fig. 6.3 The system behavior for without control ($u_1 = 0$ and $u_2 = 0$) and with control ($u_1 = 0.5$ and $u_2 = 0.025$)

infected CD4$^+$T cells initially increase because at the initial stage, CTL responses are very low. As CTL responses increase, it acts against the infected and latently infected cells and thus infected and latently infected CD4$^+$T cell population are reduced. We have also observed the dual effect of RTI and IL-2 (Fig. 6.3). In this case the infected, latently infected CD4$^+$T cells and CTL response behavior are same as like as the above mentioned cases. But the uninfected CD4$^+$T cell population increases more rapidly rather than the previous cases. In all these cases, the control (in the form of drug dose) is fixed and it is time independent. Now, our main aim is to find out the optimal strategy for the treatment to minimize the side effect as well as to minimize the cost of the treatment. On that outlook, we have used the control as a time dependent function. Numerically, we have tried to see the optimal control problem (6.1) and (6.2) as a two-point boundary value problem and we choose $t_i = 0$ and $t_f = 100$. Here we also assume that at $t_f = 100$, the treatment will be stopped. We have solved the optimality of the system by making the changes of the variable $\tau = t/t_f$ and transferring the interval $[0, 1]$. Here τ represents the step size, which is used for better strategy with a line search method and maximize the reduction of performance measure. We have chosen $t_f = 1 + \Delta t_f$ and initially $t_f = 1$. We also assume that $\Delta t_f = 1$ and our desired value of $t_f = 100$. The solutions (homotopy path) are displayed in Fig 6.4.

In Fig. 6.4, we see that the RTI starts with basic value 0.0001765 and the RTI treatment reaches to its maximum value 0.001017 in between 60 and 61 days, after that it is progressively decreased and at the end of the treatment, it reaches to zero. Furthermore for IL-2 inhibitor, we observe that the treatment starts with an initial value 0.3357 and it reaches its maximum value 0.9882 within 60–61 days as similar as RTI and again, it bit by bit reduces to zero at the end of the treatment. We have also observed that the optimal drug treatment makes a positive effect on a sharp rise

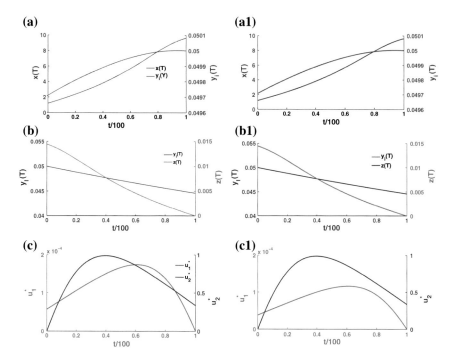

Fig. 6.4   The system behavior for the optimal treatment schedule of the control variable $u_1(t)$ and $u_2(t)$ for $P = 10$, $Q = 18$ (*left panel*) and for $P = 15$ and $Q = 18$ (*right panel*)

of uninfected CD4$^+$T cells and a gradual reduction of infected and latently infected CD4$^+$T cells. Furthermore, due to the optimal drug therapy, the CTL responses also enhance during the treatment period. Here we take two different cases, i.e., for $P = 10$, $Q = 18$ and for $P = 15$, $Q = 18$. In both the cases, the system has same behavior. However, the most interesting thing is that when R increases from 10 to 15, the RTIs attain the lower maximum value compared to the previous cases. Thus treatment against HIV, drugs like RTIs along with IL-2 are more effective compared to RTI or IL-2 treatment separately.

6.2   The Optimal Control Problem on Chemotherapy for (3.15)

In this section, our main object is to minimize the infected CD4$^+$T cell population as well as minimize the systemic cost of drug treatment. In order to that we formulate an optimal control problem. We also want to maximize the level of healthy CD4$^+$T cells. So in the model (3.15), we use a control variable $u(t)$. It represents the drug dose satisfying $0 \leq u(t) \leq 1$. Here $u(t) = 1$ represents the maximal use of chemotherapy and $u(t) = 0$ represents no treatment.

We choose our control class measurable function defined on $[t_0, t_f]$ with the condition $0 \leq u(t) \leq 1$, i.e., $U := \{u(t)|u(t)$ is measurable, $0 \leq u(t) \leq 1, t \in [t_0, t_f]\}$.

Based on the above assumptions, the optimal control problem is formulated as,

$$\text{Minimize} J[u] = \int_{t_0}^{t_f} [Py(t) + Ru^2(t) - Qx(t)]dt, \qquad (6.10)$$

subject to the state system,

$$\begin{aligned}
\dot{x} &= \lambda + px(1 - \frac{x}{T_m}) - d_1x - (1 - u(t))\beta_1xy, \\
\dot{y} &= (1 - u(t))\beta_1xy - d_2y - \beta_2yz_d - \beta_3yz_e, \\
\dot{z}_d &= sy - d_3z_d, \\
\dot{z}_e &= cxyz_e - d_4z_e.
\end{aligned} \qquad (6.11)$$

The parameters $P \geq 0$, $Q \geq 0$ and $R \geq 0$ represent the weight constants on the benefit of cost. Our main aim is to find out the optimal control variable u^* satisfying $J(u^*) = \min\limits_{0 \leq u(t) \leq 1} J(u)$.

6.2.1 Existence Condition of an Optimal Control

The boundedness of the system (6.11) for the finite time interval $[t_0, t_f]$ proves the existence of an optimal control. Since the state system (6.11) has bounded coefficient, hence we can say that the set of controls and corresponding state variables are nonempty. The control set U is convex. The right-hand side [14] of the state system (6.11) is bounded by a linear function. The integral of the cost functional $Py(t) + Ru^2(t) - Qx(t)$ is clearly convexed on U. Let there exists $c_1 > 0, c_2 > 0$ and $\eta > 1$ satisfying, $Py(t) + Ru^2(t) - Qx(t) \geq c_1|\varepsilon|^{\eta} - c_2$.

As the state system is bounded, the existence of the optimal control problem is established.

6.2.2 Characterization of an Optimal Control

Since the existence of the optimal control problem (6.10) and (6.11) are established, then Pontrygin's Minimum Principle is used to derive necessary conditions on the optimal control [2].

We define the Lagrangian in the following form as follows:

$$
\begin{aligned}
L(x, y, z_d, z_e, u, \xi_1, \xi_2, \xi_3, \xi_4) =\ & Py(t) + Ru^2(t) - Qx(t) \\
& + \xi_1[\lambda + px(1 - \frac{x}{T_m}) - d_1 x - (1 - u(t))\beta_1 xy] \\
& + \xi_2[(1 - u(t))\beta_1 xy - d_2 y - \beta_2 yz_d - \beta_3 yz_e] \\
& + \xi_3(sy - d_3 z_d) + \xi_4(cxyz_e - d_4 z_e) \\
& - w_1(t)u(t) - w_2(t)(1 - u(t)),
\end{aligned} \tag{6.12}
$$

where $w_1(t) \geq 0$ and $w_2(t) \geq 0$ are the penalty multipliers satisfying that at an optimal control $u^*(t)$, $w_1(t)u(t) = 0$ and $w_2(t)(1 - u(t)) = 0$.

The existence condition for the adjoint variables are given by,

$$
\begin{aligned}
\dot{\xi_1} &= -\frac{\partial L}{\partial x} = -[-Q + \xi_1\{p(1 - \frac{2x}{T_m}) - d_1 - (1 - u(t))\beta_1 y\} \\
& \quad + \xi_2(1 - u(t))\beta_1 y + \xi_4 cyz_e], \\
\dot{\xi_2} &= -\frac{\partial L}{\partial y} = -[P - \xi_1(1 - u(t))\beta_1 x + \xi_2\{(1 - u(t))\beta_1 x - d_2 \\
& \quad - \beta_2 z_d - \beta_3 z_e\} + \xi_3 s + \xi_4 cxz_e], \\
\dot{\xi_3} &= -\frac{\partial L}{\partial z_d} = -[-\xi_2\beta_2 y - \xi_3 d_3], \\
\dot{\xi_4} &= -\frac{\partial L}{\partial z_e} = -[-\xi_2\beta_3 y + \xi_4(cxy - d_4)],
\end{aligned} \tag{6.13}
$$

where $\xi(t_f) = 0$ for $i = 1,2,3,4$ are the transversality conditions. The Lagrangian is minimized at the optimal u^*, so the Lagrangian with respect to u at u^* is zero. Now,

$$
\begin{aligned}
L = &[-\xi_1(1 - u(t))\beta_1 xy + \xi_2(1 - u(t))\beta_1 xy - w_1(t)u(t) \\
& - w_2(t)(1 - u(t)) + Ru^2(t)] + \text{terms without } u(t).
\end{aligned} \tag{6.14}
$$

According to "Pontryagin's Minimum Principle," the unrestricted optimal control u^* satisfies $\frac{\partial L}{\partial u} = 0$ at $u = u^*$.

Therefore, $u^*(t) = \frac{(\xi_2 - \xi_1)\beta_1 xy + w_1(t) - w_2(t)}{2R}$.

To determine the explicit expression for the optimal control without w_1 and w_2 including the boundary condition, we consider the following three cases:

(i) For the set $\{t | 0 < u^*(t) < 1\}$, we have $w_1(t) = w_2(t) = 0$, hence the optimal control is: $u^*(t) = \frac{(\xi_2 - \xi_1)\beta_1 xy}{2R}$.

(ii) For the set $\{t | u^*(t) = 1\}$, we have $w_1(t) = 0$, hence $u^*(t) = 1 = \frac{(\xi_2 - \xi_1)\beta_1 xy - w_2}{2R}$.
Again since $w_2(t) > 0 \Rightarrow \frac{(\xi_2 - \xi_1)\beta_1 xy - w_2}{2R} \geq 1$.

(iii) For the set $\{t | u^*(t) = 0\}$, we have $w_2(t) = 0$.

Hence the optimal control is $0 = u^*(t) = \frac{(\xi_2 - \xi_1)\beta_1 xy + w_1}{2R}$.

Therefore $w_1(t) \geq 0 \Rightarrow \frac{(\xi_2 - \xi_1)\beta_1 xy}{2R} \leq 0$.

Combining these three cases, the optimal control is characterized as,

$$u^*(t) = \max\{0, \min(1, \frac{(\xi_2 - \xi_1)\beta_1 xy}{2R})\}. \tag{6.15}$$

If $\xi_1 > \xi_2$ for some t, then $u^*(t) \neq 1$.

Thus, $0 \leq u^*(t) \leq 1$ for such t means treatment should be administered. Hence we have the following proposition:

Proposition 6.2.1 *An optimal control u^* for system (6.11) maximizing the objective function (6.10) is characterized by (6.15).*

Thus, we find the optimal control $u^*(t)$ for (6.11) is computed with (6.13) together with (6.15). Here we have only treated the case $\delta \leq u(t) < 1, \delta > 0$ implies that chemotherapy never completely stopped viral replication.

6.2.3 Numerical Solutions of the Model Equations

In this section, we illustrate without control and with control model numerically. In the numerical simulation, we assume $x(0) = 100$, $y(t) = 50$, $z_d(t) = 2$, $z_e(t) = 5$ and the units are mm^{-3}. For the numerical illustration of the optimal control problem (6.10) and (6.11), we assume $t_f = 1$ and can be used as an initial consideration. We solve the optimality system by making the changes of the variable $\tau = t/t_f$ and transferring the interval $[0, 1]$. Here τ represents the step size, which is used for better strategy with a line search method. It will maximize the reduction of performance measure. We choose $t_f = 1 + \Delta t_f$ and initially $t_f = 1$. We also assume that $\Delta t_f = 0.1$ and our desired value of $t_f = 100$. The solutions are displayed in Figs. 6.5, 6.6, and 6.7. For different weight factors $R = 10$ and $R = 15$, numerical solution for optimal treatment strategy has been generated (Fig. 6.5). The optimal system is solved using "The Steepest Descent Method." Here we also measure minimum cost $J = 8.3442$ for $R = 10$ and $J = 10.7764$ for $R = 15$. Thus for less value of R, the optimal therapy $u^*(t)$ will achieve (Table 6.2).

Fig. 6.5 The optimal control schedule for the system (6.11)

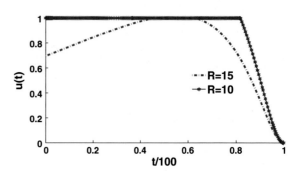

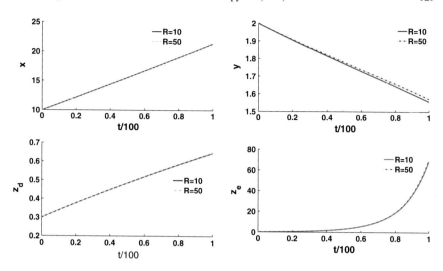

Fig. 6.6 The system behavior for optimal treatment, when final time $t_f = 1$, keeping all other parameter as in Table 6.2

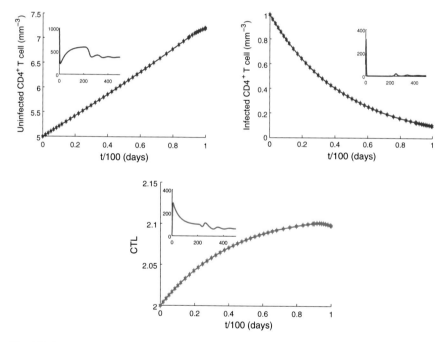

Fig. 6.7 The system behavior for the optimal treatment schedule of the control variables $u_1(t)$ and $u_2(t)$. *Inset* The system behavior in the absence of treatment

Table 6.2 Parameters used in the models (6.10) and (6.11)

Parameter	λ	p	T_m	d_1	β_1	d_2	β_2	d_3	β_3	s	c	d_4
Default value	10	0.03	1500	0.01	0.002	0.24	0.001	0.001	0.2	0.02	0.2	0.02

All parameter values are taken from [7, 9, 10, 15]

Here, it should be emphasized from collecting data that the limit of the drug dose is near about 500 days [16]. However, from our numerical solution (Fig. 6.5), if it is used for less than 50 days, the result for best treatment is to be appeared, but if it is introduced for more than 50 days, the worst condition will appear in spite of a better one. Once infection is low, the immune response is not required at high levels and this is why it too drops off. We see that when the immune response is high, less medication is needed to control and regulate infection. Our optimal treatment reduces the period of time, while the immune response of the uninfected T cells takes over. So from mathematical calculation and numerical simulation, we may infer that interruption of drug therapy is needed to allow for rebuilding the immune system.

6.3   The Optimal Control Problem for the System (4.1)

In this section, our main aim is to minimize the cost as well as the infected CD4$^+$T cell count and maximize the uninfected CD4$^+$T cells for the system (4.1). Thus, we construct the optimal control problem, where the state system is,

$$\dot{x} = \lambda + \frac{s_1}{k + y^m} - d_1 x - \beta(1 - \eta_1 u_1(t))xy + \eta_2 u_2(t)x,$$
$$\dot{y} = \beta(1 - \eta_1 u_1(t))xy - d_2 y - pyz,$$
$$\dot{z} = sy + \frac{s_2}{k + y^m} - d_3 z + \eta_3 u_2(t)z, \tag{6.16}$$

and the control function is defined as,

$$J(u_1, u_2) = \int_{t_i}^{t_f} [Pu_1^2(t) + Qu_2^2(t) - x^2 + y^2]dt. \tag{6.17}$$

The control functions $u_1(t)$ and $u_2(t)$ represent the percentage effect of viral suppression (RTIs) and immune boosting drug (IL-2). The parameters P and Q are the balancing cost factors due to scales and importance of the system of the objective function. These are the cost of per unit of RTI and IL-2 respectively.

Here we have considered that RTI reduces the infection rate by $(1 - \eta_1 u_1(t))$, where η_1 is the drug effectiveness and $u_1(t)$ is the control input doses of RTI. We have also considered the enhancement of uninfected CD4$^+$T cells and CTL responses through IL-2 treatment, defined by $\eta_2 u_2(t)$ and $\eta_3 u_2(t)$ respectively. Here $u_2(t)$ denotes control input of IL-2 treatment and η_2 and η_3 are the drug effectiveness of IL-2 for uninfected CD4$^+$T cells and CTL population respectively.

In this problem, we are seeking the optimal control pair $(u_1{}^*(t), u_2{}^*(t))$ such that

$$J(u_1{}^*, u_2{}^*) = \min\{J(u_1, u_2) : (u_1, u_2) \in U\},$$

where U is the control set defined by $U = \{u : (u_1, u_2) : u_1, u_2$ are the Lebesgue measurable, $0 \le u_1(t) \le 1, 0 \le u_2(t) \le 1, t \in [t_i, t_f]\}$.

To determine the optimal control $u_1^*(t)$ and $u_2^*(t)$, we use the "Pontryagin's Minimum Principle" [4]. To solve the problem we use the Hamiltonian given by,

$$H = Pu_1{}^2(t) + Qu_2{}^2(t) - x^2 + y^2 + \xi_1\{\lambda + \frac{s_1}{k + y^m} - d_1 x - \beta(1 - \eta_1 u_1(t))xy$$

$$+ \eta_2 u_2(t)x\}\xi_2\{\beta(1 - \eta_1 u_1(t))xy - d_2 y - pyz\} + \xi_3\{sy + \frac{s_2}{k + y^m} - d_3 z$$

$$+ \eta_3 u_2(t)z\} + w_{11}u_1(t) + w_{12}(1 - u_1(t)) + w_{21}u_2(t) + w_{22}(1 - u_2(t)).$$

$$(6.18)$$

Here $w_{11}(t)$, $w_{12}(t)$, $w_{21}(t)$ and $w_{22}(t)$ are penalty multipliers ensuring that $u_1(t)$ and $u_2(t)$ remain bounded between 0 and 1. We also have $w_{11}u_1^*(t) = w_{21}(1 - u_1^*(t)) = w_{21}u_2^*(t) = w_{22}(1 - u_2^*(t)) = 0$. The $\xi_i(t)$, $i = 1, 2, 3$ are the adjoint system together with state system, determine our optimality system.

By using the "Pontryagin's Minimum Principle" and the existence conditions for the optimal control theory [4], we obtain the following theorem:

Theorem 6.3.1 *The objective cost function $J(u_1, u_2)$ over U is minimum for the optimal control $u^* = (u_1^*, u_2^*)$ corresponding to the interior equilibrium (x^*, y^*, z^*), satisfying the state system variables together with adjoint system.*

Proof Using Pontryagin's Minimum Principle [4], the unconstrained optimal control variable $u_1^*(t)$ and $u_2^*(t)$ satisfy,

$$\frac{\partial H}{\partial u_1^*} = \frac{\partial H}{\partial u_2^*} = 0. \qquad (6.19)$$

$$H = [Pu_1{}^2(t) - \xi_1(1 - \eta_1 u_1(t))\beta xy + \xi_2(1 - \eta_1 u_1(t))\beta xy + w_{11}u_1(t)$$

$$+ w_{12}(1 - u_1(t))] + [Qu_2{}^2(t) + \xi_1 \eta_2 u_2 x + \xi_3 \eta_3 u_2 z + w_{21}u_2(t) + w_{22}(1 - u_2(t))]$$

$$+ \text{other terms without } u_1(t) \text{ and } u_2(t). \qquad (6.20)$$

Then we obtain $\frac{\partial H}{\partial u_i}$ for u_i^*, $(i = 1, 2)$ and equating to zero we get,

$$\frac{\partial H}{\partial u_1^*} = 2Pu_1^* + \beta xy\eta_1(\xi_1 - \xi_2) + w_{11} - w_{12} = 0,$$

$$\frac{\partial H}{\partial u_2^*} = 2Qu_2^* + \xi_1 \eta_2 x + \xi_3 \eta_3 z + w_{21} - w_{22} = 0.$$

Then we have,

$$
u_1{}^*(t) = \frac{\beta_1 x y \eta_1 (\xi_2 - \xi_1) + w_{12} - w_{11}}{2P},
$$

$$
u_2{}^*(t) = \frac{-(\xi_1 \eta_2 x + \xi_3 \eta_3 z) + w_{22} - w_{21}}{2Q}. \tag{6.21}
$$

Since the standard control is bounded, thus we conclude for the control $u_1^*(t)$:

$$
u_1^*(t) = \begin{cases} 0, & \frac{\beta_1 x y \eta_1 (\xi_2 - \xi_1)}{2P} \le 0; \\ \frac{\beta_1 x y \eta_1 (\xi_2 - \xi_1)}{2P}, & 0 < \frac{\beta_1 x y \eta_1 (\xi_2 - \xi_1)}{2P} < 1; \\ 1 & \frac{\beta_1 x y \eta_1 (\xi_2 - \xi_1)}{2P} \ge 1. \end{cases}
$$

The compact form of $u_1^*(t)$ is given by,

$$
u_1^*(t) = \max\{\min\{1, \frac{\beta_1 x y \eta_1 (\xi_2 - \xi_1)}{2P}\}, 0\}. \tag{6.22}
$$

In similar way, we get the compact form of $u_2^*(t)$ as,

$$
u_2^*(t) = \max\{\min\{1, \frac{-(\xi_1 \eta_2 x + \xi_3 \eta_3 z)}{2Q}\}, 0\}. \tag{6.23}
$$

The optimality of the system consists of the state system, the adjoint system, the initial conditions and the transversality conditions together with the relations as given in (6.22) and (6.23).

Utilizing (6.22) and (6.23), the optimality system characterizes the optimal control

$$
\dot{x} = \lambda + \frac{s_1}{k + y^m} - d_1 x - \beta(1 - \eta_1 u_1^*(t)) x y + \eta_2 u_2^*(t) x,
$$

$$
\dot{y} = \beta(1 - \eta_1 u_1^*(t)) x y - d_2 y - p y z,
$$

$$
\dot{z} = s y + \frac{s_2}{k + y^m} - d_3 z + \eta_3 u_2^*(t) z,
$$

$$
\frac{d\xi_1}{dt} = 2x + \xi_1\{d_1 + (1 - \eta_1 u_1^*(t))\beta y - \eta_2 u_2^*(t)\}
$$
$$
\quad - \xi_2 (1 - \eta_1 u_1^*(t))\beta y,
$$

$$
\frac{d\xi_2}{dt} = -2y + \xi_1\{(1 - \eta_1 u_1^*(t))\beta x + \frac{s_1 m y^{m-1}}{(k + y^m)^2}\}
$$
$$
\quad - \xi_2 (1 - \eta_1 u_1^*(t))\beta x - \xi_3\{s - \frac{s_2 m y^{m-1}}{(k + y^m)^2}\},
$$

$$
\frac{d\xi_3}{dt} = -\xi_2 p y + \xi_3 (d_3 - \eta_3 u_2^*(t)), \tag{6.24}
$$

satisfying $\xi_i(t_f) = 0$, $(i = 1, 2, 3)$ and $x(t_i) = x_0$, $y(t_i) = y_0$, $z(t_i) = z_0$.

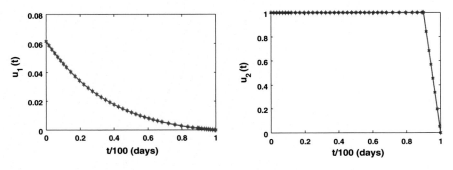

Fig. 6.8 The optimal control schedule for the system (6.16) and (6.17)

Table 6.3 Parameters used in the models (6.10) and (6.11)

Parameter	λ	d_1	β	d_2	p	s	d_3
Default value	10	$0.007 - 0.1$	0.02	0.2	0.002	0.2	0.1

All parameter values are taken from [7, 9, 10, 15]

6.3.1 Numerical Analysis

If insets of Fig. 6.7 are considered, the negative feedback effect of virus load on intensity of CTL response is very clear. During primary stage of infection when the viral load is low, CTL population is quite high, which declines when the viral count starts to increase. From Fig. 6.8 it is evident that IL-2 needs to be continued for a longer duration and given at a higher initial value compared to HAART. If the observations from Figs. 6.7 and 6.8 are combined together, it can be concluded that keeping IL-2 dose constant through time interval of treatment it reduces the dose requirement of HAART with successful enhancement of CTL and CD4$^+$T cell population and consequent decline in count of infected cells. This maximizes treatment benefit with respect to the incidence of side effects and cost. Thus if a sufficient immune response can be maintained through administration of immunomodulants such as IL-2, low drug treatment schedule with HAART can be achieved (Table 6.3).

6.4 The Optimal Control Problem of the System (4.5)

In this section, our main aim is to minimize the cost as well as minimize the infected CD4$^+$T cells and maximize the uninfected CD4$^+$T cells. In this section, we study the optimal control problem for the stable system, keeping m below 4. Thus we construct the optimal control problem where the state system is,

$$\dot{x} = \lambda + \frac{s_1}{k^m + y^m} - d_1 x - \beta(1 - u_1(t))xy + u_2(t)x,$$

$$\dot{y} = \beta(1 - u_1(t))xy - d_2 y - pyz,$$

$$\dot{z} = sy + \frac{s_2}{k^m + y^m} - d_3 z + u_2(t)z, \qquad (6.25)$$

and the control function is defined as,

$$J(u_1, u_2) = \int_{t_i}^{t_f} [Pu_1^2 + Qu_2^2 - x^2 + y^2]dt. \qquad (6.26)$$

Here the control functions $u_1(t)$ and $u_2(t)$ represent the percentage effect of RTI and IL-2 for interaction of T cells with virus. The parameters P and Q represent respectively the weight factors on the benefit of the cost of RTI and IL-2 therapy. Here the control functions $u_1(t)$ and $u_2(t)$ are bounded [3]. The control $u_1(t)$ represents the efficacy of the drug therapy in inhibiting the reverse transcription, i.e., blocking new infection. The control $u_2(t)$ represents the efficacy of IL-2 therapy. In this problem, we are seeking the optimal control pair $(u_1{}^*, u_2{}^*)$ such that $J(u_1{}^*, u_2{}^*) = \min\{J(u_1, u_2) : (u_1, u_2) \in U\}$, where U is the control set defined by $U = \{u = (u_1, u_2) : u_1, u_2$ are the measurable, $0 \le u_1(t) \le 1, 0 \le u_2(t) \le 1, t \in [t_i, t_f]\}$.

To determine the optimal control $u_1{}^*$ and $u_2{}^*$, we use the "Pontryagin Minimum Principle" [4]. To solve the problem, we use the Hamiltonian given by

$$H = Pu_1^2 + Qu_2^2 - x^2 + y^2 + \xi_1\{\lambda + \frac{s_1}{k^m + y^m} - d_1 x - \beta(1 - u_1(t))xy + u_2(t)x\}$$

$$+ \xi_2\{\beta(1 - u_1(t))xy - d_2 y - pyz\} + \xi_3\{sy + \frac{s_2}{k^m + y^m} - d_3 z + u_2(t)z\}. \quad (6.27)$$

Using the "Pontryagin's Minimum Principle" and the existence conditions for the optimal control theory [4], we obtain the Proposition.

Proposition 6.4.1 *The objective cost function $J(u_1, u_2)$ over U is minimum for the optimal control $u^* = (u_1{}^*, u_2{}^*)$ corresponding to the interior equilibrium (x^*, y^*, z^*). Also there exists adjoint functions ξ_1, ξ_2, ξ_3 satisfying the Eq. (6.32).*

Proof Using Pontryagin's Minimum Principle [4], the unconstrained optimal control variable $u_1{}^*$ and $u_2{}^*$ satisfy,

$$\frac{\partial H}{\partial u_1{}^*} = \frac{\partial H}{\partial u_2{}^*} = 0. \qquad (6.28)$$

$$H = [Pu_1{}^2 - \xi_1(1 - u_1(t))\beta xy + \xi_2(1 - u_1(t))\beta xy] + [Qu_2{}^2 + \xi_1 u_2 x + \xi_3 4 u_2 z]$$
$$+ \text{ other terms without } u_1 \text{ and } u_2, \tag{6.29}$$

then we get $\frac{\partial H}{\partial u_i{}^*}$ for $u_i{}^*$, $(i = 1, 2)$ and equating to zero we get,

$$\frac{\partial H}{\partial u_1{}^*} = 2Pu_1{}^* + \beta xy(\xi_1 - \xi_2) = 0 \text{ and } \frac{\partial H}{\partial u_2{}^*} = 2Qu_2{}^* + \xi_1 x + \xi_3 z = 0.$$

Thus we have,

$$u_1{}^* = \frac{\beta_1 xy(\xi_2 - \xi_1)}{2P} \text{ and } u_2{}^* = -\frac{(\xi_1 x + \xi_3 z)}{2Q}. \tag{6.30}$$

Then according to the standard control arguments, we can conclude for u_1^*:

$$u_1^* = \begin{cases} 0, & \frac{\beta_1 xy(\xi_2 - \xi_1)}{2P} \leq 0; \\ \frac{\beta_1 xy(\xi_2 - \xi_1)}{2P}, & 0 < \frac{\beta_1 xy(\xi_2 - \xi_1)}{2P} < 1; \\ 1 & \frac{\beta_1 xy(\xi_2 - \xi_1)}{2P} \geq 1. \end{cases}$$

In compact form, we can rewrite $u_1^*(t) = \min\{1, \frac{\beta xy(\xi_2 - \xi_1)}{2P}\}$. Similarly, for $u_2^*(t)$ we have the compact form $u_2^* = \min\{1, -\frac{\xi_1 x + \xi_3 z}{2Q}\}$.

According to Pontryagin's Minimum Principle [4],

$$\frac{d\xi}{dt} = -\frac{\partial H}{\partial x} \quad \text{and} \tag{6.31}$$

$$H(x(t), u^*(t), \xi(t), t) = \min_{u \in U} H(x(t), u(t), \xi(t), t).$$

The above equations are the necessary conditions satisfying the optimal control $u(t)$ and the adjoint variables. The adjoint system in our problem becomes,

$\frac{d\xi_1}{dt} = -\frac{\partial H}{\partial x}$, $\frac{d\xi_2}{dt} = -\frac{\partial H}{\partial y}$, $\frac{d\xi_3}{dt} = -\frac{\partial H}{\partial z}$. Taking the partial derivative of H in (6.27) we get,

$$\frac{d\xi_1}{dt} = 2x + \xi_1\{d_1 + (1 - u_1(t))\beta y - u_2(t)\} - \xi_2(1 - u_1(t))\beta y,$$

$$\frac{d\xi_2}{dt} = -2y + \xi_1\{(1 - u_1(t))\beta x + \frac{s_1 m y^{m-1}}{(k^m + y^m)^2}\} - \xi_2(1 - u_1(t))\beta x - \xi_3\{s - \frac{s_2 m y^{m-1}}{(k^m + y^m)^2}\},$$

$$\frac{d\xi_3}{dt} = -\xi_2 py + \xi_3(d_3 - u_2(t)). \tag{6.32}$$

We have analyzed the optimality of the system, which consists of the state system, the adjoint system together with the initial conditions and the transversality conditions. The transversability conditions are given at final time t_f by $\xi_i(t_f) = 0$, $i = 1, 2, 3$.

Using the state system (6.25) and the adjoint system (6.32), we can conclude that the objective function (6.26) will be minimized for u_1^*, and u_2^* (6.30) with the initial conditions are $x(0) = x_0$, $y(0) = y_0$, $z(0) = z_0$ and the transversality condition $\xi_i(t_f) = 0$, $i = 1, 2, 3$.

6.4.1 Numerical Analysis

Here we choose the initial conditions for the state variables as $x_0 = 5$, $y_0 = 1$, $z_0 = 2$ [5] and we have also used the parameters given in Table 6.4. The solutions are displayed in Fig. 6.9 (Table 6.4). Here we have plotted the trajectories of the state variables and the optimal control variables for different values of the cost in the form of weight factors. From the numerical study, we have observed that if the weight factor of IL-2 be increased, then the treatment control u_2 will remain at upper bound for short period of time in comparison with respect to the earlier case. Also it is

Table 6.4 Variable values at the final time of treatment for the optimal control input with different sets of cost

P	Q	x_f	y_f	z_f	u_1^*	u_2^*
10	10	22.52	0.8076	6.724	0.04251 at 0.1023	1 up to 0.9798

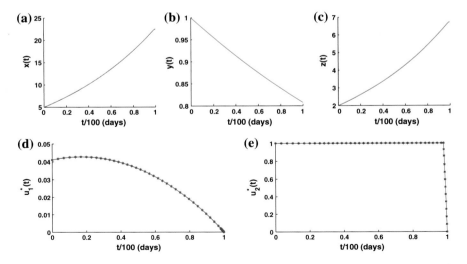

Fig. 6.9 Optimal trajectories of the state variables and control variables for $P = 10$ and $Q = 10$

observed that the treatment control u_1 takes more time to reach at its maximum value (which is comparatively lower to previous one). We have also observed that the weight factor of HAART does not make any significant impact on the system due to the negative feedback effect. Thus the combination of drug therapeutic treatment (HAART and IL-2) is more effective in the presence of negative feedback effect.

6.5   Optimization of the System (3.22)

The control class is chosen to be measurable functions defined on $[t_i, t_f]$ with the condition $0 \leq u(t) \leq 1$, i.e.,
$U := \{u(t) | u(t) \text{ is Lebesgue measurable with values between 0 and 1, } 0 \leq u(t) \leq 1, \ t \in [t_i, t_f]\}$.

Together with the state system (3.22), we consider an optimal control problem with the cost function given by,

$$J(u) = \int_{t_i}^{t_f} [y^2(t) - x^2(t) + Bu^2(t)]dt. \qquad (6.33)$$

The term $Bu^2(t)$ represents systematic cost of the drug treatment. We try to find out an optimal control function u^* such that $J(u^*) = min\{J(u)|u \in U\}$. We apply "Pontryagin Minimum Principle" [1] to determine the specific optimal control u^* of our problem. To do this, we start by defining Hamiltonian. The Hamiltonian for our problem consists of the integrand of the cost functional and the right-hand side of the state equations through the adjoint variables $\lambda_1, \lambda_2, \lambda_3$ and penalty multipliers ω_1 and ω_2 joining the control constraints.

The Hamiltonian is defined as follows:

$$H(x, y, z, u, \lambda_1, \lambda_2, \lambda_3) = y^2 - x^2 + Bu^2 + \lambda_1 \left[\lambda + rx \left(1 - \frac{x}{c} \right) - (1 - u)\beta xy - dx \right]$$
$$+ \lambda_2 \left[(1 - u)\beta xy - ay - \rho yz \right] + \lambda_3 \left[\frac{ky^n}{1 + ky^n} - bz \right] - w_1 u - w_2 (1 - u), \qquad (6.34)$$

where $w_1(t) \geq 0$ and $w_2(t) \geq 0$ are the penalty multipliers which can make sure that $u(t)$ remains bounded between 0 and 1, satisfying $w_1(t)u(t) = 0$ and $w_2(t)(1 - u(t)) = 0$.

Theorem 6.1 *Corresponding to the state system and optimal control u^*, there exists adjoint variables $\lambda_1, \lambda_2,$ and λ_3 satisfying,*

$$\lambda_1' = -\frac{\partial H}{\partial x} = 2x - \lambda_1\left[r - \frac{2rx}{c} - (1-u)\beta y - d\right] - \lambda_2(1-u)\beta y,$$

$$\lambda_2' = -\frac{\partial H}{\partial y} = -2y + \lambda_1(1-u)\beta x - \lambda_2\left[(1-u)\beta x - a - \rho z\right]$$

$$-\lambda_3\left[\frac{nky^{n-1}}{1+ky^n} - \frac{nk^2 y^{2n-1}}{(1+ky^n)^2}\right],$$

$$\lambda_3' = -\frac{\partial H}{\partial z} = \lambda_2 \rho y + \lambda_3 b, \tag{6.35}$$

with the transversality conditions $\omega_i(t_f) = 0$ for $i = 1, 2, 3$. Further $u^*(t)$ is represented by,

$$u^*(t) = \max\left\{0, \min\left\{1, \left(\frac{(\lambda_2 - \lambda_1)\beta xy}{2B}\right)^+\right\}\right\}. \tag{6.36}$$

Proof We differentiate the Hamiltonian H with respect to x, y and z respectively, and then the adjoint system can be written as $\lambda_1' = -\frac{\partial H}{\partial x}$, $\lambda_2' = -\frac{\partial H}{\partial y}$ and $\lambda_3' = -\frac{\partial H}{\partial z}$. To find the optimal control, we differentiate the Hamiltonian H with respect to u, which gives $\frac{\partial H}{\partial u} = \lambda_1 \beta xy - \lambda_2 \beta xy - w_1 + w_2 + 2Bu = 0$ at u^*. Solving we get $u^*(t) = \frac{(\lambda_2 - \lambda_1)\beta xy + w_1 - w_2}{2B}$. To determine an explicit expression for the optimal control without w_1 and w_2, we consider the following three cases:

(i) for $t|0 < u^*(t) < 1$, we have $w_1(t) = w_2(t) = 0$, hence the optimal control is: $u^*(t) = \frac{(\lambda_2 - \lambda_1)\beta xy}{2B}$,

(ii) for $t|u^*(t) = 1$, we have $w_1 = 0$, hence $u^*(t) = 1 = \frac{(\lambda_2 - \lambda_1)\beta xy - w_2}{2B}$, which implies $w_2(t) \geq 0, 1 \leq \frac{(\lambda_2 - \lambda_1)\beta xy}{2B}$,

(iii) for $t|u^*(t) = 0$, we have $w_1(t) \geq 0, w_2(t) = 0$, hence the optimal control is: $u^*(t) = 0 = \frac{(\lambda_2 - \lambda_1)\beta xy + w_1}{2B}$.

Therefore, $w_1(t) \geq 0$ implies that $(\lambda_2 - \lambda_1)\beta xy \leq 0$.
Hence,

$$\left(\frac{(\lambda_2 - \lambda_1)\beta xy}{2B}\right)^+ = 0 = u^*(t).$$

Combining these three cases, the optimal control is characterized as,

$$u^*(t) = \max\left\{0, \min\left\{1, \left(\frac{(\lambda_2 - \lambda_1)\beta xy}{2B}\right)^+\right\}\right\}.$$

We have shown that the optimality system consists of state system with initial conditions, the adjoint system, transversality conditions, and optimality conditions.

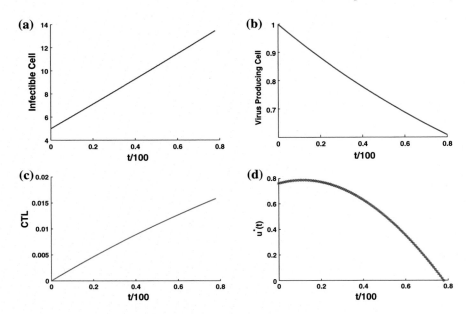

Fig. 6.10 This figure shows the system behavior for the optimal treatment schedule of the control variable $u(t)$ for $B = 2$

6.5.1 Numerical Simulation

Figure 6.10d represents the control u^* for RTI drug. The RTI drug is administered at nearly full level for 20 days approximately and after that it is reduced to zero at 78 days. In Fig. 6.10a we see that during the treatment period, the infectible T cells increase almost linearly and the CTL responses also increase almost linearly (see Fig. 6.10c), whereas the virus producing cells linearly decrease (see Fig. 6.10b).

6.6 The Optimal Control Problem (2.11)

In this section, our main object is to minimize the infected CD4$^+$T cell population as well as minimize the systematic cost of drug treatment. Here we formulate an optimal control problem. We also want to maximize the level of healthy CD4$^+$T cells. So in our basic model (2.11), we use control variables $u_1(t)$ and $u_2(t)$ that represent the drug dose satisfying $0 \leq u_1(t) \leq 1$ and $0 \leq u_2(t) \leq 1$. Here $u_1(t) = 1$, $u_2(t) = 1$ represent the maximal use of chemotherapy and $u_1(t) = 0$ and $u_2(t) = 0$ represent no treatment. Here we consider that the RTI reduces the infection rate by

$(1 - \eta_1 u_1)$, where η_1 represents the drug efficacy and u_1 is the control input doses of the drug RTI. We also consider the enrichment of uninfected T cells and CTL responses through IL-2 treatment are given by $\eta_2 u_2$ and $\eta_3 u_2$, where u_2 as a control input of IL-2 treatment and η_2 and η_3 are the drug efficacy of IL-2 for uninfected T cells and precursor CTL responses. In this section, our main aim is to minimize the cost as well as minimize the infected CD4$^+$T cells and maximize the uninfected CD4$^+$T cell. Thus we construct the optimal control problem where the state system is,

$$\begin{aligned}
\dot{x} &= \lambda - dx - \beta(1 - \eta_1 u_1(t))xy + (1 - \eta_2 u_2(t))\gamma x, \\
\dot{y} &= \beta(1 - \eta_1 u_1(t))xy - ay - pyz, \\
\dot{w} &= cxyw - cq_1 yw - bw + (1 - \eta_3 u_2(t))\gamma_1 w, \\
\dot{z} &= cq_2 yw - hz,
\end{aligned} \tag{6.37}$$

and the control function is defined as (Fig. 6.11),

$$J(u_1, u_2) = \int_{t_0}^{t_f} [y(t) - x(t) - w(t) + Ru_1^2 + Pu_2^2]dt, \tag{6.38}$$

where the parameters R and P respectively represent the weight on the benefit of the cost.

Here the control function $u_1(t)$ and $u_2(t)$ are bounded and Lebesgue integrable function [3].

In this problem, we are seeking the optimal control pair $(u_1{}^*, u_2{}^*)$ such that

$$J(u_1{}^*, u_2{}^*) = \min\{J(u_1, u_2) : (u_1, u_2) \in U\}.$$

Here U is the control set defined by $U = \{u = (u_1, u_2) : u_1, u_2 \text{ are the measurable,} \; 0 \le u_1(t) \le 1, \; 0 \le u_2(t) \le 1, \; t \in [t_0, t_f]\}$.

To determine the optimal control $u_1{}^*$ and $u_2{}^*$, we use the "Pontryagin Minimum Principle" [1]. To solve the problem, we use the Hamiltonian given by,

$$\begin{aligned}
H =\; & y(t) - x(t) - w(t) + Ru_1^2 + Pu_2^2 + \xi_1\{\lambda - dx - \beta(1 - \eta_1 u_1(t))xy + (1 - \eta_2 u_2(t))\gamma x\} \\
& + \xi_2\{\beta(1 - \eta_1 u_1(t))xy - ay - pyz\} + \xi_3\{cxyw - cq_1 yw - bw + (1 - \eta_3 u_2(t))\gamma_1 w\} \\
& + \xi_4\{cq_2 yw - hz\}.
\end{aligned} \tag{6.39}$$

Using the "Pontryagin Minimum Principle" and the existence conditions for the optimal control theory [4], we obtain the following theorem:

Theorem 6.6.1 *The objective cost function $J(u_1, u_2)$ over U is minimum for the optimal control $u^* = (u_1{}^*, u_2{}^*)$ corresponding to the interior equilibrium (x^*, y^*, w^*, z^*). Also there exists adjoint functions ξ_1, ξ_2, ξ_3 and ξ_4 satisfying the Eqs. (6.39) and (6.40).*

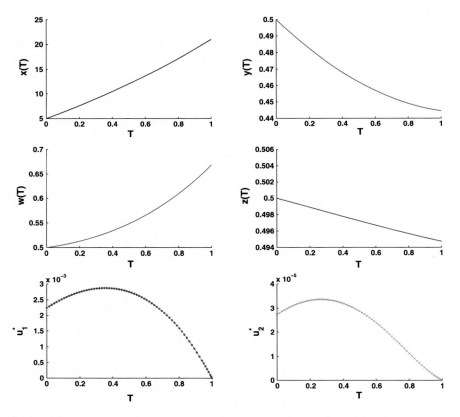

Fig. 6.11 This figure shows the system behavior for the optimal treatment schedule of the control variables $u_1(t)$ and $u_2(t)$ for $R = 10$, $P = 50$, and $\eta_1 = 0.1$, $\eta_2 = 0.1$, $\eta_3 = 0.01$

Proof Using Pontryagin Minimum principle [4], the unconstrained optimal control variables u_1^* and u_2^* satisfy,

$$\frac{\partial H}{\partial u_1^*} = \frac{\partial H}{\partial u_2^*} = 0. \tag{6.40}$$

$$H = [Ru_1^2 - \xi_1(1 - \eta_1 u_1(t))\beta xy + \xi_2(1 - \eta_1 u_1(t))\beta xy] + [Pu_2^2 + \xi_1(1 - \eta_2 u_2)\gamma x$$
$$+ \xi_3(1 - \eta_3 u_2)\gamma_1 w] + \text{other terms without } u_1 \text{ and } u_2, \tag{6.41}$$

then we obtain $\frac{\partial H}{\partial u_i^*}$ for u_i^* ($i = 1, 2$), and hence equating to zero becomes,

$$\frac{\partial H}{\partial u_1^*} = 2Ru_1^* + \eta_1 \beta x^* y^* (\xi_1 - \xi_2) = 0,$$

$$\frac{\partial H}{\partial u_2^*} = 2Pu_2^* - \xi_1 \eta_2 \gamma x^* - \xi_3 \eta_3 \gamma_1 z^* = 0.$$

Thus we obtain the optimal control u_1^* and u_2^* corresponding to the interior equilibrium (x^*, y^*, w^*, z^*) as,

$$u_1^* = \frac{\beta \eta_1 x^* y^* (\xi_2 - \xi_1)}{2R} \quad \text{and} \quad u_2^* = \frac{(\xi_1 \eta_2 \gamma x^* + \xi_3 \eta_3 \gamma_1 z^*)}{2P}. \tag{6.42}$$

According to "Pontryagin Minimum Principle" [1] we know that

$$\frac{d\xi}{dt} = -\frac{\partial H}{\partial x}, \tag{6.43}$$

and

$$H(x(t), u^*(t), \xi(t), t) = \min_{u \in U} H(x(t), u(t), \xi(t), t). \tag{6.44}$$

The above equations are the necessary conditions satisfying the optimal control $u(t)$ and again for the system (6.37, 6.38) the adjoint equations are,

$$\frac{d\xi_1}{dt} = -\frac{\partial H}{\partial x}, \quad \frac{d\xi_2}{dt} = -\frac{\partial H}{\partial y}, \quad \frac{d\xi_3}{dt} = -\frac{\partial H}{\partial w}, \quad \frac{d\xi_4}{dt} = -\frac{\partial H}{\partial z}.$$

Taking the partial derivative of H we get,

$$\frac{d\xi_1}{dt} = 1 + \xi_1 \{d_1 + (1 - \eta_1 u_1(t))\beta y - (1 - \eta_2 u_2(t))\gamma\} - \xi_2(1 - \eta_1 u_1(t))\beta y - \xi_3 cyw,$$

$$\frac{d\xi_2}{dt} = -1 + \xi_1(1 - \eta_1 u_1(t))\beta x - \xi_2\{(1 - \eta_1 u_1(t))\beta x - a - pz\} - \xi_3\{cxw - cq_1 w\} - \xi_4 cq_2 w,$$

$$\frac{d\xi_3}{dt} = 1 - \xi_3\{cxy - cq_1 y - b + (1 - \eta_3 u_2(t))\gamma_1\} - \xi_4 cq_2 y,$$

$$\frac{d\xi_4}{dt} = \xi_2 py + \xi_4 h. \tag{6.45}$$

Hence the optimality of the system consists of the state system along with the adjoint system. Also it depends on the initial conditions and the transversality conditions, which satisfy $\xi_i(t_f) = 0$ and $x(0) = x_0$, $y(0) = y_0$, $w(0) = w_0$ and $z(0) = z_0$.

6.6.1 Discussion

From optimal control studies, several interesting results have been obtained. As weight factors (R and P) increase from 0.1 to 10 and 0.5 to 50 respectively, there is a significant increase in the uninfected cell population with very little effect on the count of infected cells. The increase in the weight factors does not produce proportionate increase in the precursor CTL population and increase in effectiveness of HAART or IL-2 as denoted by η, which is not manifested by remarkable change in precursor CTL number. The CTL effector population is found to decrease in all the cases. Thus, optimal control approach will help in designing an innovative cost-effective safe therapeutic regimen of HAART and IL-2, where the uninfected cell population will be enhanced with simultaneous decrease in the infected cell population. Moreover, successful immune reconstitution can also be achieved with increase in precursor CTL population. Mathematical modeling of viral dynamics thus enables maximization of therapeutic outcome even in case of multiple therapies with specific goal of reversal of immunity impairment.

From our above discussion of the results, it is clear that adoption of optimal control strategy in optimization of therapeutic regime of combination therapy of HAART and IL-2 is found to be really satisfactory in terms of enhancing the life expectancy of HIV-afflicted patients by improving the uninfected T cell count.

6.7 Optimal Control Strategy

Here our main object is to minimize the level of infected $CD4^+T$ cell population and maximize the level of healthy $CD4^+T$ cells. From this understanding and the relation mentioned in (2.12), the immune response is also to be increased with the simultaneous occurrence of uninfected T cell population and depletion of infected cell population. Also, it is our object to keep in mind that cost as measured in terms of chemotherapy strength, which is a combination of time and efficacy is to made as low as possible. To reach our intention for cost effective better treatment of a HIV patient, we thus formulate an optimal control problem and hence in the model equation (2.12), we include a control variable $\eta(t)$, which represents the drug dose and satisfying $0 \le \eta(t) \le 1$. Here $\eta(t) = 1$ represents the maximal use of chemotherapy and $\eta(t) = 0$, which signifies no treatment. Since the control reduces viral replication rate, we multiply our infectivity term βxy by $(1 - \eta)$. It should be noted here that both cellular infection rate and viral production rate are represented by the same term β, so the drug may be represented either a Protease Inhibitor or a Reverse Transcriptase Inhibitor drug [17].

We choose our control class measurable function defined on $[t_{start}, t_{final}]$ with the condition $0 \leq \eta(t) \leq 1$, i.e., $U := \{\eta(t) | \eta(t) \text{ is measurable}, 0 \leq \eta(t) \leq 1, t \in [t_{start}, t_{final}]\}$. Based on the above assumptions, an optimal control problem is to be formulated as:

$$\frac{dx}{dt} = \lambda - cx - (1 - \eta)\beta xy,$$

$$\frac{dy}{dt} = (1 - \eta)\beta xy - by - \frac{\alpha yz}{a + y},$$

$$\frac{dz}{dt} = \frac{uyz}{a + y} - dz. \tag{6.46}$$

Define the objective function as,

$$J(\eta) = \int_{t_{start}}^{t_{final}} [y(t) + R\eta(t)^2 - x(t)]dt . \tag{6.47}$$

Here the parameter $R \geq 0$ represents the desired 'weight constant' on the benefit and cost. Thus it is imperative to characterize the optimal control u^*, satisfying $min_{0 \leq \eta \leq 1} J(\eta) = J(\eta^*)$.

We consider the Hamiltonian is as follows,

$$H(x, y, z, \eta, \omega_1, \omega_2, \omega_3) = y(t) + R\eta^2 - x + \omega_1[\lambda - cx - (1 - \eta)\beta xy],$$

$$+ \omega_2[(1 - \eta)\beta xy - by - \frac{\alpha yz}{a + y}] + \omega_3[\frac{uyz}{a + y} - dz - v_1\eta - v_2(1 - \eta)], \quad (6.48)$$

where $v_1(t) \geq 0$ and $v_2(t) \geq 0$ are the penalty multipliers, satisfying $v_1(t)\eta(t) = 0$ and $v_2(t)(1 - \eta(t)) = 0$. Thus, the Minimum Principal give the existence of adjoint variables satisfying,

$$\omega_1' = -\frac{\partial H}{\partial x} = 1 + \omega_1 c + \omega_1(1 - \eta)\beta y - \omega_2(1 - \eta)\beta y,$$

$$\omega_2' = \frac{\partial H}{\partial y} = -1 + \omega_1(1 - \eta)\beta x - \omega_2(1 - \eta)\beta x + \omega_2 b$$

$$+ \frac{\omega_2 \alpha z}{a + y} - \frac{\omega_2 \alpha yz}{(a + y)^2} - \frac{\omega_3 uz}{a + y} + \frac{\omega_3 uyz}{(a + y)^2},$$

$$\omega_3' = -\frac{\partial H}{\partial z} = \frac{\omega_2 \alpha y}{a + y} - \frac{\omega_3 uy}{a + y} + \omega_3 d, \tag{6.49}$$

where $\omega_i(t_{final}) = 0$ for $i = 1, 2, 3$ are the transversality condition. The Hamiltonian is minimized with respect to η at the optimal η^*, so the derivative of the Hamiltonian with respect to η at η^* is zero. Again,

$$H = [-\omega_1(1 - \eta)\beta xy + \omega_2(1 - \eta)\beta xy - v_1\eta - v_2(1 - \eta) + R\eta^2]$$
$$+ \text{terms without } \eta(t). \quad (6.50)$$

Differentiating this expression for H with respect to η gives,

$$\frac{\partial H}{\partial \eta} = \omega_1\beta xy - \omega_2\beta xy - v_1 + v_2 + 2R\eta = 0 \quad \text{at } \eta^*. \quad (6.51)$$

Solving for the optimal control yields,

$$\eta^*(t) = \frac{(\omega_2 - \omega_1)\beta xy + v_1 - v_2}{2R}. \quad (6.52)$$

Now we try to find out $\eta^*(t)$, using different penalty multipliers in three different cases:

(i) For $t|0 \le \eta^*(t) \le 1$, we have $v_1(t) = v_2(t) = 0$. Hence the optimal control is,

$$\eta^*(t) = \frac{(\omega_2 - \omega_1)\beta xy}{2R}.$$

(ii) For $t|\eta^*(t) = 1$, we have $v_1 = 0$. Hence, $\eta^*(t) = 1 = \frac{(\omega_2-\omega_1)\beta xy-v_2}{2R}$, which implies

$$v_2(t) > 0, 1 \le \frac{(\omega_2 - \omega_1)\beta xy}{2R}.$$

(iii) For $t|\eta^*(t) = 0$, we have $v_1(t) \ge 0$, $v_2(t) = 0$. Hence the optimal control is,

$$\eta^*(t) = 0 = \frac{(\omega_2 - \omega_1)\beta xy + v_1}{2R}.$$

Therefore, $v_1(t) \ge 0$ implies that $(\omega_2 - \omega_1)\beta xy \le 0$, hence

$$(\frac{(\omega_2 - \omega_1)\beta xy - v_2}{2R})^+ = 0 = \eta^*(t).$$

Combining the above three cases, the optimal control is characterized as,

$$\eta^*(t) = \max\{0, \min\{1, (\frac{(\omega_2 - \omega_1)\beta xy - v_2}{2R})^+\}\}. \quad (6.53)$$

If $(\omega_2 - \omega_1) < 0$ for some t, then $\eta^*(t) \ne 1$. So we can conclude that $0 < \eta^*(t) < 1$ for those t, where infected population should be administered.

Theorem 6.7.1 *An optimal control η^* for Eq. (6.46) which minimizes the objective function (6.47) is characterized by (6.53), where the notation $\varepsilon^+ = \max(\varepsilon, 0)$. We could have only treated the case $\varepsilon \leq \eta \leq 1, \varepsilon > 0$, which would say that the drug never completely stop infected cell population's reproduction.*

6.7.1 Numerical Experiment of Optimal Control Strategy

For the numerical illustration of the optimal control problem (6.46) and (6.47), we assume $t_{\text{final}} = 1$, which can be used as an initial guess. Using different combinations of weight factors R, one can generate several treatment schedules for various time periods. The solution is displayed in Figs. 6.12 and 6.13.

In Fig. 6.12, we have plotted two different cases using $R = 3$ and $R = 4$. All cases are during the treatment period. Here we see that uninfected T cell population increases proportionately with the weight function R and the infected T cell decreases disproportionate to R. The Lymphocyte cell population increases initially but after $t_f = 0.76$, it decreases with proportionate to R.

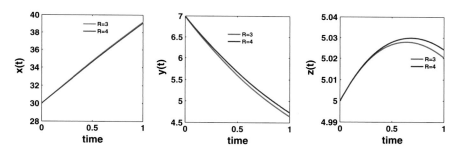

Fig. 6.12 The system behavior for optimal treatment, when final time $t_f = 1$. We take the value of $a = 9$

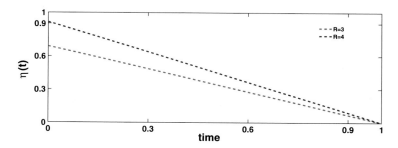

Fig. 6.13 Optimal treatment schedule with different R

Figure 6.13 shows that for different weight factors $R = 3$ and $R = 4$, numerical solution for optimal treatment strategy has been generated. The optimal system is solved using "The Steepest Descent Method". In this figure we see that as R increases, the control variable $\eta(t)$ decreases proportionately. Thus the percentage of chemotherapy is proportionally related to the weight factor R. Here we also measure minimum cost $J = 0.4124$ for $R = 3$ and $J = 0.1530$ for $R = 4$. Thus for greater value of R, the optimal therapy $\eta^*(t)$ will achieve. Here it should be emphasized from collecting data that the maximum duration of the drug dose of chemotherapy is near about 500 days. But from our numerical solution (Fig. 6.12), if it is used for less than 76 days the result for best treatment is to be appeared, but if it is introduced for more than 76 days then it would reflect a negative result and CTL population decreases. So from mathematical calculation and numerical simulation we may infer that interruption of drug therapy is needed to allow for reconstruction of the immune system.

6.8 The Optimal Control Problem in case of Recovery of Infected Cells in HIV Model

Here our main object is to minimize the level of infected CD4$^+$T cell population and maximize the level of healthy $CD4^+T$ cells. From this understanding we wish to make relation where the immune response is also to be increased with the simultaneous occurrence of uninfected T cell population and depletion of infected cell population. Also, it is our object to keep in mind that cost as measured in terms of chemotherapy strength, which is a combination of time and efficacy is made as low as possible. To reach our intention for cost effective better treatment of a HIV patient, we thus formulate an optimal control problem for the model proposed by Zhou et al. [18], we use a control variable $u(t)$, which represents the drug dose at a time t satisfying $0 \le u(t) \le 1$. Here $u(t)$ represents control input with values normalized in between 0 and 1 [5]. Also $u(t) = 1$ represents the maximal use of chemotherapy and $u(t) = 0$, which signifies no treatment. Since the control reduces viral replication rate, so after using control the infection rate becomes $\beta(1 - u(t))$. It should be noted here that both cellular infection rate and viral production rate are represented by the same term β, so that the drug RTI may be introduced for better treatment [17].

Based on the above assumptions, our optimal control problem corresponding to the equation [18] would be,

$$\frac{dT}{dt} = s - d_1 T + aT(1 - \frac{T}{T_{\max}}) - (1 - u(t))\beta TV + \rho I,$$
$$\frac{dI}{dt} = (1 - u(t))\beta TV - \delta I - \rho I,$$
$$\frac{dV}{dt} = qI - cV, \tag{6.54}$$

satisfying the initial conditions $T(0) = T_0$, $I(0) = I_0$ and $V(0) = V_0$.
    The objective function is defined as,

$$J(u) = \int_{t_i}^{t_f} [I(t) - T(t) + Ru^2(t)]dt. \tag{6.55}$$

The parameter $R(>0)$ is the weight constant on the benefit of the cost. The benefit is based on the minimization of cost and infected cell count together with maximization of uninfected T cell count. Now, we want to find the optimal control u^* such that $J(u^*) = \min\{J(u) : u \in U\}$ subject to the system of ODE's (6.54), where U is the admissible control set defined by
    $U = \{u(t) : u(t) \text{ is measurable}, 0 \leq u(t) \leq 1, t \in [t_i, t_f]\}$ is the control set.
    To determine the optimal control, we use the "Pontryagin Minimum Principle" [1].
    Let us consider the Hamiltonian given by,

$$\begin{aligned}
H = {} & I(t) - T(t) + Ru^2(t) + \xi_1\{s - d_1 T + aT(1 - \frac{T}{T_{max}}) \\
& - (1 - u(t))\beta TV + \rho I\} + \xi_2\{(1 - u(t))\beta TV - \delta I - \rho I\} \\
& + \xi_3(qI - cV),
\end{aligned} \tag{6.56}$$

where ξ_1, ξ_2 and ξ_3 are the adjoint variables.

Theorem 6.8.1 *If the given optimal control u^* and the solutions T^*, I^*, V^* of the corresponding state system (6.54) minimize $J(u)$ over U, then there exists the adjoint variables ξ_1, ξ_2 and ξ_3, which satisfy the following equations,*

$$\frac{d\xi_1}{dt} = -\left[-1 + \xi_1\{-d_1 + a(1 - \frac{2T}{T_{max}}) - \beta V(1 - u(t))\} + \xi_2 \beta V (1 - u(t))\right],$$

$$\frac{d\xi_2}{dt} = -[1 + \xi_1 \rho - \xi_2(\delta + \rho) + \xi_3 q],$$

$$\frac{d\xi_3}{dt} = -[-\xi_1 \beta T(1 - u(t)) + \xi_2 \beta T(1 - u(t)) - \xi_3 c],$$

along with the transversality condition $\xi_i(t_f) = 0$ for $i = 1, 2, 3$. Further, u^* is represented by $u^* = \max\{0, \min\{\frac{\beta TV(\xi_2 - \xi_1)}{2R}, 1\}\}$.

Proof The Hamiltonian (6.56) can be written as,

$$\begin{aligned}
H = {} & Ru^2(t) + \xi_1 u(t)\beta TV - \xi_2 u(t)\beta TV \\
& + \text{other term without } u(t).
\end{aligned} \tag{6.57}$$

According to the "Pontryagin Minimum Principal" [1], the unconstrained optimal control variable u^* satisfies,

$$\frac{\partial H}{\partial u^*} = 0, \qquad (6.58)$$

i.e., $\frac{\partial H}{\partial u^*} = 2Ru^*(t) + \beta T V (\xi_1 - \xi_2) = 0.$
   Hence, we obtain from the above expression,

$$u^* = \frac{\beta T V (\xi_2 - \xi_1)}{2R}. \qquad (6.59)$$

By applying the standard control arguments involving the bounds on the control [20], we conclude for u,

$$u^* = \begin{cases} 0, & \frac{\beta T V (\xi_2 - \xi_1)}{2R} \leq 0; \\ \frac{\beta T V (\xi_2 - \xi_1)}{2R}, & 0 < \frac{\beta T V (\xi_2 - \xi_1)}{2R} < 1; \\ 1, & \frac{\beta T V (\xi_2 - \xi_1)}{2R} \geq 1. \end{cases}$$

To ensure the positivity of u^*, we use the following notation $u^+ = \max(s, 0)$ [21]. Therefore in compact notation,

$$u^* = \min\{\frac{\beta T V (\xi_2 - \xi_1)^+}{2R}, 1\}. \qquad (6.60)$$

Now, according to the "Pontryagin Minimum Principal" [1], we can write

$$\frac{d\xi_i}{dt} = -\frac{\partial H}{\partial x}, i = 1, 2, 3, \qquad (6.61)$$

where $x \equiv (T, I, V)$ and the necessary conditions satisfying the optimal control $u(t)^*$ are,

$$H\left(x(t), u^*(t), \xi_i(t), t\right) = \min_{u \in U} H\left(x(t), u(t), \xi_i(t), t\right), i = 1, 2, 3. \quad (6.62)$$

So, the adjoint equations corresponding to the system (6.57) are,

$$\frac{d\xi_1}{dt} = -\frac{\partial H}{\partial T}, \quad \frac{d\xi_2}{dt} = -\frac{\partial H}{\partial I}, \quad \frac{d\xi_3}{dt} = -\frac{\partial H}{\partial V}.$$

Therefore,

$$\frac{d\xi_1}{dt} = -\left[-1 + \xi_1\left(-d_1 + a(1 - \frac{2T}{T_{max}}) - \beta V(1 - u(t))\right) + \xi_2 \beta V(1 - u(t))\right],$$

$$\frac{d\xi_2}{dt} = -\left[1 + \xi_1 \rho - \xi_2(\delta + \rho) + \xi_3 q\right],$$

$$\frac{d\xi_3}{dt} = -\left[-\xi_1 \beta T(1 - u(t)) + \xi_2 \beta T(1 - u(t)) - \xi_3 c\right], \qquad (6.63)$$

where $\xi_i(t_f) = 0, i = 1, 2, 3$ are transversality conditions and $T(0) = T_0, I(0) = I_0,$ $V(0) = V_0$ are initial conditions.

6.8.1 Numerical Simulation and Discussion

The dynamics of the disease progression of HIV/AIDS along with control measures are analyzed using numerical methods. Here we build a computer simulation technique with a particular choice of parameter values, which are reasonably realistic. The parameter values, considered here, are shown in Table 6.5. We find the quasi-stationary distribution and the expected time to extinction of the infected $CD4^+T$ cells. Here we also consider the initial conditions for the state variables T, I, and V along with the transversality conditions for the adjoint variables ξ_i, $i = 1, 2, 3$, where $\xi_i(t_f) = 0, i = 1, 2, 3$. Here we analyze the effect of change of recovery rate

Table 6.5 A hypothetical set of parameter values

Parameters	Values
s	$5 \, day^{-1} \, mm^{-3}$ [18]
d_1	$0.01 \, day^{-1}$ [18]
a	$0.5 \, day^{-1}$ [18]
a_1	0.30 [19]
b_1	0.015 [19]
a_2	0.02 [19]
b_2	0.001 [19]
β	$0.0002 \, mm^{-3}$ [18]
k	$\beta q/c$ [18]
ρ	$0.01 \, day^{-1}$ [18]
δ	$1 \, day^{-1}$ [18]
q	$800 \, day^{-1} \, mm^{-3}$ [18]
c	$5 \, day^{-1}$ [18]

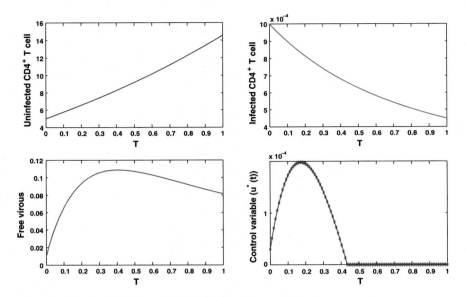

Fig. 6.14 The behavior of the state system and the control variable $u(t)$ of the optimality system (6.54) and (6.55) for $\rho = 0.01$, R = 0.1

of the infected individuals in the presence of the control theoretic approach and find the expected time to extinction of the infected CD4$^+$T cells.

Figure 6.14 represents the behavior of the disease dynamics for control theoretic approach using control variable $u(t)$ for different values of recovery rate ($\rho = 0.01$, $\rho = 0.1$, and $\rho = 0.5$). From these figures, it can be inferred that if we increase the recovery rate from $\rho = 0.01$ to $\rho = 0.1$ and then $\rho = 0.5$ with the control approach of the drug adherence, then the number of infected CD4$^+$T cell decreases remarkably. This is due to the fact that with increasing recovery, antigen in the infected cells is being exhausted. In this way, after reaching a certain level, the number of virus also decreases. In the above figures, it has also been observed that by increasing the recovery rate of the infected cells, the infected CD4$^+$T cell decreases but it does not extinct. With this idea, we perform the numerical simulation for finding out the marginal distribution of the number of infected CD4$^+$T cells and its time to extinction.

References

1. Pontryagin, L.S., Boltyanskii, V.G., Gamkrelidze, R.V., Mishchenko, E.F.: Mathematical Theory of Optimal Processes, vol. 4, p. 115. Gordon and Breach Science Publishers (1986)
2. Kamien, M., Schwartz, N.L.: Dynamic Optimization, 2nd edn. North Holland (1991)
3. Swan, G.M.: Application of Optimal Control Theory in Biomedicine, vol. 135 (1984)

4. Fleming, W., Rishel, R.: Deterministic and Stochastic Optimal Controls, Chap. 3, Theorem 4.1, pp. 545–562. Springer, New York (1975)
5. Culshaw, R.V., Rawn, S., Spiteri, R.J.: Optimal HIV treatment by maximising immuno response. J. Math. Biol. **48**, 545–562 (2004)
6. Kirschner, D.E., Webb, G.F.: Immunotherapy of HIV-1 infection. J. Biol. Syst. **6**(1), 71–83 (1998)
7. Perelson, A.S., Krischner, D.E., De-Boer, R.: Dynamics of HIV infection of CD4 T cells. Math. Biosc. **114**(81–125), 118 (1993)
8. Roy, P.K., Chatterjee A.N.: Effect of HAART on CTL mediated immune cells: an optimal control theoretic approach. In: Ao, S.I., Gelman, L. (eds.) Electrical Engineering and Applied Computing, vol. 90, pp. 595–607, Springer, New York (2011)
9. Bonhoeffer, S., Coffin, J.M., Nowak, M.A.: Human immunodeficiency virus drug therapy and virus load. J. Virol. **71**, 3275–3278 (1997)
10. Wodarz, D., Nowak, M.A.: Specific therapy regimes could lead to long-term immunological control to HIV. Proc. Natl. Acad. Sci. USA **96**(25), 14464–14469 (1999)
11. Wodarz, D., May, R.M., Nowak, M.A.: The role of antigen-independent persistence of memory cytotoxic T lymphocytes. Int. Immunol. V **12**(A), 467–477 (2000)
12. Wodarz, D., Nowak, M.A.: The mathematical models of HIV pathogenesis and treatment. BioEssays **24**, 1178–1187 (2002)
13. Gray, D.: T cell and B cell memory are short lived in the absence of antigen. J. Cell. Biochem. Suppl. **13A**, CO10 (1989)
14. Lukas, D.L.: Differential Equation: Classical to Controlled. Mathematical in Science and Engneering. Academic Press, New York (1982)
15. Perelson, A.S., Neuman, A.U., Markowitz, J.M.,Leonard, Ho, D.D.: HIV 1 dynamics in vivo: viron clearance rate, infected cell life span, and viral generation time. Science **271**, 1582–1586 (1996)
16. Kirschner, D.E., Webb, G.F.: Resistance, remission, and qualitative difference in HIV chemotherapy. Emerg. Infect. Dis. **3**(3), 273–283 (1997)
17. Perelson, A.S., Nelson, P.W.: Mathematical analysis of HIV-1 dynamics in Vivo. SIAM Rev. **41**(3–41), 122 (1999)
18. Zhou, X., Song, X., Shi, X.: A differential equation model of HIV infection of $CD4^+T$ cells with cure rate. J. Math. Anal. Appl. **342**, 1342–1355 (2008)
19. Matis, J.H., Kiffe, T.R.: On the cumulants of population size for the stochastic power law logistic model. Theor. Popul. Biol. **53**, 16–29 (1998)
20. Joshi, H.R.: Optimal control of an HIV immunology model. Optimal Control Appl. Methods **23**, 199–213 (2002)
21. Kirschner, D., Lenhart, S., Serbin, S.: Optimal control of the chemotherapy of HIV. J. Math. Biol. **35**, 775–792 (1997)

Chapter 7
Perfect Drug Adherence

Abstract The effect of perfect drug adherence towards controlling the disease HIV/AIDS is discussed through impulsive differential equations. Here, we have assumed the model with both drugs RTIs and IL-2 that are taken at each impulse $t = t_k (k = 1, 2, 3, \ldots)$ and $t = T_l (l = 1, 2, 3, \ldots)$ respectively. Furthermore, we have considered that the effects of the drugs are instantaneous. However, the system endures a prompt change in the state. Here, we have considered the mathematical models including combination of drug therapies (T-20 and IL-2). Here, we have mainly studied the dynamical behavior of the system in the presence of drug. Using impulsive differential equations, dosing interval and threshold value of dosages can be obtained more precisely. We also have determined the threshold value of the drug dosage and the dosing interval for which the disease-free equilibrium remains stable.

Keywords Impulsive effect · Chemokine analog · Perfect adherence · Dimeric complex · Chemokine coreceptor

7.1 Drug Therapy with Perfect Adherence in Explicit Form

In this section, we have assumed the model (2.6) with both the drugs RTIs and IL-2 taken at each impulse at $t = t_k (k = 1, 2, 3, \ldots)$ and $t = T_l (l = 1, 2, 3, \ldots)$ respectively. Furthermore, we consider that the effects of the drugs are instantaneous. The solutions are assumed to be continuous at $t \neq t_k$ and $t \neq T_l$. However, the system endures an instantaneous change in the state. Thus, the model equations in the explicit form is given below:

$$\dot{x} = \lambda - d_1 x - \beta_1 x y_i - p_1 x R + q_1 x I,$$
$$\dot{y}_i = \beta_1 x y_i - d_2 y_i - \beta_2 y_i z - p_2 y_i R + \delta y_l I,$$
$$\dot{y}_l = v y_i - d_3 y_l - \delta y_l I,$$
$$\dot{z} = s y_i - d_4 z + q_2 z I,$$
$$x' = p_1 x R - d_1 x',$$
$$\dot{y}_i' = p_2 y_i R - d_2 y_i'. \tag{7.1}$$

© Springer Science+Business Media Singapore 2015
P.K. Roy, *Mathematical Models for Therapeutic Approaches
to Control HIV Disease Transmission*, Industrial and Applied Mathematics,
DOI 10.1007/978-981-287-852-6_7

Here, $R(t)$ and $I(t)$ denote the drug concentration in the plasma of RTIs and IL-2 respectively. The dynamics of the drugs are given by,

$$\frac{dR}{dt} = -d_R R, \quad t \neq t_k, \qquad \frac{dI}{dt} = -d_I I, \quad t \neq T_l. \tag{7.2}$$

The impulsive conditions are given by,

$$\Delta R = R_d, \quad t = t_k, \qquad \Delta I = I_d, \quad t = T_l. \tag{7.3}$$

The drug RTIs acts on the uninfected CD4$^+$T cells and infected CD4$^+$T cells at rates p_1 and p_2 respectively. Furthermore, q_1 and q_2 are the rates of drug effect of IL-2 at which the uninfected CD4$^+$T cells and CTL responses are activated respectively. Here, d_R and d_I are the decay rates of drug RTI and IL-2. Furthermore, dosages ΔR_k and ΔI_l are the drug dosages respectively taken at $t = t_k$ and $t = T_l$.

Thus we can write $\Delta R_k = R(t_k^+) - R(t_k^-)$ for RTI and $\Delta I_l = I(T_l^+) - I(T_l^-)$ for IL-2.

Since here our discussion is only based upon drug perfect adherence, we assume $\Delta R_k = R_d$ and $\Delta I_l = I_d$ for all time $t = t_k$ and $t = T_l$. If no dose is taken, then we can say that $\Delta R_k = 0$ and $\Delta I_l = 0$.

7.1.1 Analysis of the Model

In the absence of the drug, the system has two equilibria.
They are

(i) the disease-free equilibrium $(x, y_i, y_l, z, x', y_i') = (\frac{\lambda}{d_1}, 0, 0, 0, 0, 0)$ and
(ii) the endemic equilibrium $(x, y_i, y_l, z, x', y_i') = (\bar{x}, \bar{y}_i, \bar{y}_l, \bar{z}, 0, 0)$, where $\bar{x} = \frac{\lambda}{d_1 + \beta_1 \bar{y}_i}$, $\bar{y}_l = \frac{v \bar{y}_i}{d_3 + \delta}$, $\bar{z} = \frac{s \bar{y}_i}{d_4}$ and $\bar{y}_i$ is the positive root of the equation

$$s\beta_1\beta_2(d_3 + \delta)y_i^2 + [s\beta_2 d_1(d_3 + \delta) + d_2 d_4(d_3 + \delta) - v\delta d_4 \beta_1]y_i$$
$$- [d_1 d_4 v\delta + \lambda\beta_1 d_4(d_3 + \delta) - d_1 d_2 d_4(d_3 + \delta)] = 0.$$

The characteristic equation is given by,

$$(\eta + d_2)(\eta + d_1)^2(\eta + d_3)(\eta + d_4)(\eta + d_2 - \beta_1\frac{\lambda}{d_1}) = 0. \tag{7.4}$$

All eigenvalues have negative real parts if the basic reproductive ratio $= \frac{\lambda\beta_1}{d_1 d_2} < 1$.

Hence, we can conclude that the disease-free equilibrium exists in the absence of drug, if the basic reproduction number is less than unity.

We also study the drug administration in the impulsive way. However, impulsive model does not reveal any steady state [1]. Since the impulsive differential equation exposes impulsive periodic orbit, thus in this case the system will exhibit periodic orbits with discontinuity. If the drugs are present and in the absence of virus, there exists a disease-free impulsive orbit,

$$(\hat{x}, 0, 0, 0, \hat{x}', \hat{y}'_i, R^*, I^*) = (\frac{\lambda}{d_1 + p_1 R^* - q_1 I^*}, 0, 0, 0, \frac{p_1 R^* \lambda}{d_1 (d_1 + p_1 R^* - q_1 I^*)}, 0, R^*, I^*).$$

Also, there exists endemic periodic orbit $\tilde{E}^* (\tilde{x}^*, \tilde{y}_i^*, \tilde{y}_l^*, \tilde{z}^*, \tilde{x}'^*, \tilde{y}_i'^*, R^*, I^*)$, where

$$\tilde{x}^* = \frac{\lambda}{d_1 + \beta_1 \tilde{y}_i^* + p_1 R^* - q_1 I^*},$$

$$\tilde{y}_l^* = \frac{v \tilde{y}_i^*}{d_3 + \delta I^*},$$

$$\tilde{z}^* = \frac{s \tilde{y}_i^*}{d_4 - q_2 I^*},$$

$$\tilde{x}'^* = \frac{p_1 R^* \tilde{x}^*}{d_1},$$

$$\tilde{y}_i'^* = \frac{q_2 R^* \tilde{y}_i^*}{d_2}$$

and

$$s\beta_1\beta_2(d_3 + \delta I^*)\tilde{y}_i^{*2} + [s\beta_2(d_3 + \delta I^*)(d_1 + p_1 R^* - q_1 I^*) - v\delta\beta_1 I^*(d_4 - q_2 I^*) + \beta_1(d_4 - q_2 I^*)(d_3 + \delta I^*)(d_2 + p_2 R^*)]\tilde{y}_i^* - [v\delta I^*(d_4 - q_2 R^*)(d_1 + p_1 R^* - q_1 I^*) + \lambda\beta_1(d_4 - q_2 I^*)(d_3 + \delta I^*) - (d_2 + p_2 R^*)(d_4 - q_2 I^*)(d_3 + \delta I^*)(d_1 + p_1 R^* - q_1 I^*) = 0.$$

7.1.2 Dynamics of the Drug

In this section, we try to find out the general solution of the drug dynamics of

$$\frac{dR}{dt} = -d_R R, \quad \text{for } t \neq t_k, \quad t_k = k\tau, \quad k = 0, 1, 2, \ldots$$

$$R(0) = R_d, \quad \text{for } t = 0, \qquad R(t_k^+) = R(t_k^-) + \Delta R_k, \quad \text{for } t = t_k. \quad (7.5)$$

Here, R_d is the initial drug dose and $\Delta R_k \geq 0$. We also consider that the impulse time t_k is fixed and the interval between two dosages is given as $\tau \equiv t_{k+1} - t_k$. Hence, the solution for the interval $[t_k, t_{k+1}]$ is given below:

$$R(t) = R(t_k^+)e^{-d_R(t - t_k)}, \quad \text{for } t_k < t \leq t_{k+1}. \quad (7.6)$$

For $t \to t_{k+1}^-$,

$$R(t_{k+1}^-) = R(t_k^+)e^{-d_R\tau}. \quad (7.7)$$

Here, we have considered $r = e^{-d_R\tau}$. The last equation of (7.5) becomes

$$R(t_{k+1}^+) = R(t_k^+)r + \Delta R_k. \quad (7.8)$$

We have considered the initial drug dose $R(t_0^+) = R_d = \Delta R_0$. From (7.6), we get

$$R(t) = R(t_k^+)e^{-d_R(t-t_k)} = \left(\sum_{j=0}^{k} r^{k-j}\Delta R_j\right)e^{-d_R(t-t_k)},$$

where $k = \lfloor\frac{t}{\tau}\rfloor$, $t \in [t_k^+, t_{k+1}^-]$.   (7.9)

The drug concentration before the drug is taken is $R(t_k^+) = \left(\sum_{j=0}^{k} r^{k-j}\Delta R_j\right)$.

The drug concentration after the drug is taken is $R(t_{k+1}^-) = \left(\sum_{j=0}^{k} r^{k-j}\Delta R_j\right)e^{-d_R\tau}$.

Now for perfect adherence of the drug therapy $\Delta R_j = R_d$.

Hence, for $[t_k^+, t_{k+1}^-]$, the drug dynamics becomes,

$$R(t) = R_d\frac{1-r^{1+k}}{1-r}e^{-d_R(t-t_k)}, \quad \text{where } k = \lfloor\frac{t}{\tau}\rfloor, \quad t_k = k\tau.   \quad (7.10)$$

Thus for the limiting case, the drug concentration before and after one dosage is being as follows: $\lim_{k\to\infty} R(t_k^+) = \frac{R_d}{1-r}$, $\lim_{k\to\infty} R(t_{k+1}^-) = \frac{R_d r}{1-r}$, and $R(t_{k+1}^+) = \frac{R_d r}{1-r} + R_d = \frac{R_d}{1-r}$.

Here, $R(t_k^+) - \frac{R_d}{1-r} = R_d\frac{1-r^{1+k}}{1-r} - \frac{R_d}{1-r} = -R_d\frac{r^{1+k}}{1-r} < 0$.

We can conclude that the positive impulsive orbit for RTIs starts at $\frac{R_d e^{-d_R\tau}}{1-e^{-d_R\tau}}$ and ends at $\frac{R_d}{1-e^{-d_R\tau}}$.

The drug dynamics of IL-2 are given below:

$$\frac{dI}{dt} = -d_I I, \quad \text{for } t \neq T_l, \quad T_l = l\sigma, \quad l = 0, 1, 2, \ldots$$

$$I(0) = I_d, \quad \text{for } t = 0, \quad I(T_l^+) = I(T_l^-) + \Delta I_l, \quad \text{for } t = T_l,   \quad (7.11)$$

where dosage interval is $\sigma \equiv T_{l+1} - T_l$. We can derive in the similar way for IL-2 therapy, where the positive impulsive orbit starts at $\frac{I_d e^{-d_I\sigma}}{1-e^{-d_I\sigma}}$ and ends at $\frac{I_d}{1-e^{-d_I\sigma}}$. Hence, we can conclude that the impulsive periodic orbit for RTI and IL-2 satisfies the following conditions stated below:

$$\frac{R_d e^{-d_R\tau}}{1-e^{-d_R\tau}} \leq R^* \leq \frac{R_d}{1-e^{-d_R\tau}}, \quad \text{and} \quad \frac{I_d e^{-d_I\sigma}}{1-e^{-d_I\sigma}} \leq I^* \leq \frac{I_d}{1-e^{-d_I\sigma}}.   \quad (7.12)$$

Now, we have considered that σ is fixed. Our aim is to find out the relation between dosing interval and dosage of RTI.

In the presence of drugs, the basic reproduction number of the system becomes,

$$R_c = \frac{\lambda \beta_1}{(d_1 + p_1 R^* - q_1 I^*)(d_2 + p_2 R^*)}.$$

Let $\phi = \frac{e^{-d_R \tau}}{1-e^{-d_R \tau}}$ and $\psi = \frac{1}{1-e^{-d_R \tau}}$. Then for fixed σ, we get $R_d \phi \le R^* \le R_d \psi$.
For disease-free steady state, the basic reproduction number $R_c < 1$.
Then, $\frac{\lambda \beta_1}{(d_1 + p_1 R_d \phi - q_1 I^*)(d_2 + p_2 R_d \phi)} < 1$.

$$\Rightarrow p_1 p_2 \phi^2 R_d{}^2 + \{p_2(d_1 - q_1 I^*) + d_2 p_1\}\phi R_d + d_2(d_1 - q_1 I^*) - \lambda \beta_1 > 0.$$
$$\Rightarrow M_2 R_d{}^2 + M_1 R_d + M_0 > 0. \tag{7.13}$$

Also for the endemic steady state, $R_c > 1$,

$$\Rightarrow p_1 p_2 \psi^2 R_d{}^2 + \{p_2(d_1 - q_1 I^*) + d_2 p_1\}\psi R_d + d_2(d_1 - q_1 I^*) - \lambda \beta_1 < 0.$$
$$\Rightarrow N_2 R_d{}^2 + N_1 R_d + N_0 < 0. \tag{7.14}$$

Here,

$M_0 = N_0 = d_2(d_1 - q_1 I^*) - \lambda \beta_1,$
$M_1 = [p_2(d_1 - q_1 I^*) + d_2 p_1]\phi,$
$M_2 = p_1 p_2 \phi^2,$
$N_1 = [p_2(d_1 - q_1 I^*) + d_2 p_1]\psi,$
$N_2 = p_1 p_2 \psi^2.$

Now the relation between the drug dosages and the drug dose interval has been established by the following theorem.

Theorem 7.1.1 *Let $R_1 = \frac{\chi}{\phi}$ and $R_2 = \frac{\chi}{\psi}$, then there exists the following cases:*

(i) *If $R_d > R_1$, the disease-free periodic orbit ($\tilde{E}_0$) exists and the endemic periodic orbit ($\tilde{E}^*$) does not exist for $0 \le \tau < \tau_1$.*
(ii) *When the dosage satisfies $0 \le R_d < R_2$, $\tilde{E}_0$ is unstable and $\tilde{E}^*$ exists for $\tau > \tau_2$.*

Proof We have $R_d \phi \le R^* \le R_d \psi$. If $R_d > R_1$, then $\Rightarrow \chi < R^*$.

Thus $\chi = \frac{p_2(q_1 I^* - d_1) - p_1 d_2 + \sqrt{\{p_2(q_1 I^* - d_1) - p_1 d_2\}^2 + 4 p_1 p_2\{(q_1 I^* - d_1)d_2 + \lambda \beta_1\}}}{2 p_1 p_2} < R^*$, which
implies $R_c < 1$ and for this case $0 \le \tau < \tau_1 \Rightarrow \tau_1 = \frac{1}{d_R}\log(1 + \frac{R_d}{\chi})$.
If $R_d < R_2$, then $R_d < \chi(1 - e^{-d_R \tau}) \Rightarrow \frac{R_d}{1-e^{-d_R \tau}} < \chi \Rightarrow R^* < \chi$,
which implies $R_c > 1$ and in this case $\tau > \tau_2$.

Hence for $\tau = \tau_2$, $\frac{R_d}{1-e^{-d_R \tau_2}} = \chi \Rightarrow \tau_2 = -\frac{1}{d_R}\log(1 - \frac{R_d}{\chi})$.
Therefore, if drugs are taken in such a way that the dosing interval τ and dosage R_d satisfy the above relation, then the disease can be controlled theoretically.

Now, the following theorem has been established in support of the drug effect on the system.

Theorem 7.1.2 *If RTIs are taken with sufficient frequency, then $x(t) + x'(t) \to \frac{\lambda}{d_1}$ and $y_i(t) + y_i'(t) \to 0$ as $t \to \infty$, $\tau \to 0$ for any σ fixed.*

Proof For the system (7.1), we get $(x + y_i + y_l + z + x' + y_i') \le \frac{\lambda}{d_1}$. Then, $z(t) \le \frac{s\lambda}{d_1(d_4 - I^*)} \le f(t, \sigma, \tau)$. Also, $y_l(t) \le \frac{v\lambda}{d_1(d_3 + \delta I^*)}$. Again,

$$\frac{dx}{dt} + \frac{dy_i}{dt} + \frac{dy_i'}{dt} \le \lambda + [\frac{v\delta\lambda}{d_1(d_3 + \delta I^*)} + \frac{q_1\lambda}{d_1}]I^* - \frac{p_1\lambda R^*}{d_2} - d_1(x + y_i + y_i'),$$

$$\Rightarrow x(t) + y_i(t) + y_i'(t) \le \frac{\lambda}{d_1} + [\frac{v\delta\lambda}{d_1^2(d_3 + \delta I^*)} + \frac{q_1\lambda}{d_1^2}]I^* - \frac{p_1\lambda R^*}{d_1 d_2},$$

$$\Rightarrow y_i(t) + y_i'(t) \le [\frac{v\delta\lambda}{d_1^2(d_3 + \delta I^*)} + \frac{q_1\lambda}{d_1^2}]I^* - \frac{p_1\lambda R^*}{d_1 d_2} = \phi(t, \sigma, \tau),$$

$$\to 0, \text{ as } t \to \infty, \tau \to 0, \sigma \text{ fixed.} \quad (7.15)$$

Now,

$$x(t) + x'(t) > \frac{\lambda + [q_1 f_1(t, \sigma, \tau) - \beta_1\phi(t, \sigma, \tau)]\frac{\lambda}{d_1}}{d_1} \quad (7.16)$$

$$\to \frac{\lambda}{d_1}, \text{ as } t \to \infty, \tau \to 0, \sigma \text{ fixed.} \quad (7.17)$$

Remark 7.1.1 Keeping interval of the IL-2 dosage fixed, if RTIs are taken with sufficient interval, then certain amount of drug dosage may be able to sustain the immune system at pre-infection level and the infected CD4$^+$T cells are moved towards zero.

7.1.3 Numerical Simulation of the Explicit Model

In this section, we have analyzed Figs. 7.1 and 7.2. All parameters are taken from Table 7.1. We choose the initial conditions as, $x(0) = 1000$, $y_i(0) = 100$, $y_l(0) = 10$, $z(0) = 500$, $x'(0) = 0$, $y_i'(0) = 0$ and the unit of the concentration is mm^{-3}.

Figure 7.1 shows the region of stability in the presence of drug therapy. If R_d is sufficiently large and τ is sufficiently small, then E_0 is stable (upper region) and E^* does not exist. However, E_0 loses its stability and E^* exists for sufficiently small value of R_d and large value of τ (lower region). However, the accurate threshold lies between these two regions.

Figure 7.2 shows the change of concentration for each model variable with respect to time in the presence of perfect adherence. In this figure, the dosing interval is fixed

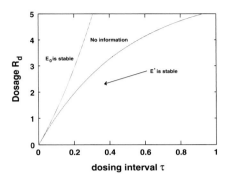

Fig. 7.1 The region of stability in the presence of drug therapy

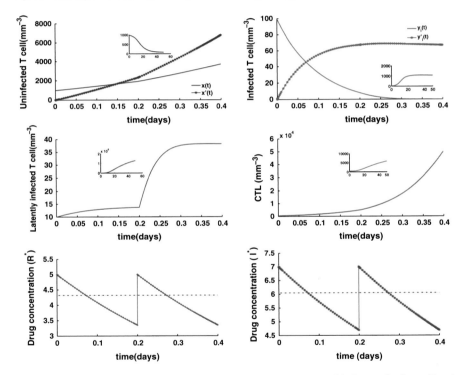

Fig. 7.2 The impulsive system behavior for the perfect adherence with $R_d = 5$, $I_d = 7$ and $\tau = \sigma = 0.2$. *Inset* Trajectories for the system in the absence of therapy

at $\tau = \sigma = 0.2$. This means that the drugs are taken five times a day. From this figure we can easily conclude that after introducing the therapy, the healthy CD4$^+$T cell population increases as time progress. Furthermore, the infected CD4$^+$T cell population moves towards extinction and the latently infected and infected CD4$^+$T cells move towards a steady-state region. During the treatment, the CTL population

Table 7.1 List of parameters for systems (7.1)–(7.3)

Parameter	d_1	d_2	d_3	d_4	β_2	δ	ν	s
Default value	0.03	0.3	0.03	0.2	0.001	0.1	0.5	0.2
Parameter	β_1	λ	p_1	p_2	q_1	q_2	d_R	d_I
Default value	0.0001	10	2	2	2	2	2	2

Units are $\text{mm}^{-3}\text{day}^{-1}$ except β_2, s, d_4, p_1, p_2, q_1, q_2, as because their units are day^{-1}

enhances comparatively with the absence of drug treatment (Inset Figure). We also numerically find the mean value of R^* and I^* from this figure, which are 4.32 and 6.048 respectively.

7.2   Enfuvirtide-IL-2 Administration for HIV-1 Treatment

In this section, we consider a mathematical model [2], including the combination of drug therapies (T-20 and IL-2). Here, we mainly study the dynamical behavior of the system in the presence of drug. Using impulsive differential equation, dosing interval and threshold value of dosages can be obtained more precisely. We also determine the threshold value of the drug dosage and the dosing interval for which the disease-free equilibrium remains stable using impulsive differential equations.

In this section, we use $T_s(t)$ to represent the concentration of uninfected CD4$^+$T cells, $T_l(t)$ denotes the concentration of latently infected CD4$^+$T cells, $T_i(t)$ denotes the concentration of actively infected CD4$^+$T cells, and $V(t)$ denotes the concentration of free HIV-1 virus. Here, $R(t)$ denotes the intracellular concentration of the fusion inhibitor, while $I(t)$ represents intracellular concentration of IL-2. We modify the mathematical model [2] and incorporate combination of drug therapies (T-20 and IL-2) into the model as given below.

7.2.1   Combining T Cell Population with Virus and Drugs

$$\dot{T}_s = \lambda - d_s T_s - \beta T_s V + \alpha T_s I,$$
$$\dot{T}_l = \beta T_s V - d_s T_l - \delta T_l I - \eta T_l,$$
$$\dot{T}_i = \eta T_l - d_i T_i + \delta T_l I,$$
$$\dot{V} = \mu T_i - d_v V - kVF. \tag{7.18}$$

Here, λ is the constant production rate of uninfected CD4$^+$T cells, which are produced from bone marrow and mature in thymus. Like any other cells, T cells are also removed due to their natural apoptosis. Here, d_s represents the natural death rate of uninfected and latently infected CD4$^+$T cells and β is the rate at which free virus infects uninfected CD4$^+$T cells. During infection process, CD4$^+$T cells remain in the latent state. Latently infected cell becomes an actively infected cell at a rate η. Natural death rate of actively infected CD4$^+$T cells is d_i. Actively infected CD4$^+$T cells produce free virus at a rate μ and d_v is the killing effect of HIV-1 by the whole immune system. The term kVF represents the effect of T-20 Inhibitor. Binding rate of T-20 with HIV virus is k. The terms $\alpha T_s I$ and $\delta T_l I$ represent the effect of IL-2 activator on uninfected CD4$^+$T cells and latently infected CD4$^+$T cells respectively. Here, α and δ are the rates of activation of uninfected CD4$^+$T cells and latently infected CD4$^+$T cells respectively through IL-2.

In this section, we have assumed that drugs are given at independent times t_k for the enfuvirtide inhibitor and s_k for the IL-2 activator (not necessarily fixed). The effect of the drugs is assumed to be instantaneous. The solutions are continuous at $t \neq t_k$ and $t \neq s_k$.

The dynamics of two drugs, F and I, are given by

$$\frac{dF}{dt} = -m_F F, \quad t \neq t_k, \quad \frac{dI}{dt} = -m_I I, \quad t \neq s_k. \qquad (7.19)$$

The impulsive conditions are,

$$\Delta F = F^i, \quad t = t_k, \quad \Delta I = I^i, \quad t = s_k. \qquad (7.20)$$

Here, m_R and m_I are the rates at which T-20 and IL-2 are cleared. F^i is the dose of fusion inhibitor (T-20) and I^i is the dose of interleukin-2 (IL-2). The system of ordinary differential equations together with two difference equations (7.19)–(7.20) represents the system of impulsive differential equations. According to impulsive theory [3], we have

$$F(t_k^+) = F(t_k^-) + F^i, \quad I(s_k^+) = I(s_k^-) + I^i. \qquad (7.21)$$

Here, we consider only the drug with perfect adherence and we assume that $\Delta F = F^i$, and $\Delta I = I^i$ for all time $t = t_k$ and $t = s_k$. If no dose is taken, then we can say that $\Delta F = 0$ and $\Delta I = 0$.

7.2.2 Analysis of the Model

In general, impulsive models do not exhibit steady states [4], but rather impulsive periodic orbits (periodic orbits with discontinuities) exist. Since the impulsive differential equation exposes impulsive periodic orbit, thus in this case the system will exhibit periodic orbits with discontinuity.

In the presence of drugs and in the absence of virus, there exists a disease-free impulsive orbit $E_1(\widehat{T}_s, 0, 0, 0, F^*, I^*) = (\frac{\lambda}{d_s - \alpha I^*}, 0, 0, 0, F^*, I^*)$. Also, there exists endemic periodic orbit $E^*(\overline{T_s^*}, \overline{T_l^*}, \overline{T_i^*}, \overline{V^*}, F^*, I^*)$, where

$$\overline{T_s^*} = \frac{\lambda}{\beta \overline{V^*} + d_s - \alpha I^*}, \qquad \overline{T_l^*} = \frac{d_i \overline{V^*}(d_v + kF^*)}{\mu(\eta + \delta I^*)}, \qquad \overline{T_i^*} = \frac{\overline{V^*}(d_v + kF^*)}{\mu},$$

$$\overline{V^*} = \frac{\lambda \beta \mu(\eta + \delta I^*) - d_i(d_v + kF^*)(d_s + \eta + \delta I^*)(d_s - \alpha I^*)}{\beta d_i(d_v + kF^*)(d_s + \eta + \delta I^*)}.$$

Here, F^* and I^* are not constant but other parameters are constant.

7.2.2.1   Basic Reproduction Number

The characteristic equation at the disease-free impulsive orbit E_1 becomes

$$(d_s - \alpha I^* + \xi)(\xi^3 + A_1 \xi^2 + A_2 \xi + A_3) = 0, \qquad (7.22)$$

where
$$
\begin{aligned}
A_1 &= d_i + d_s + d_v + \delta I^* + \eta + kF^*, \\
A_2 &= d_i d_s + d_s d_v + d_i d_v + \eta d_i + \delta I^* d_i + \eta kF^* + kF^* \delta I^* \\
&\quad + kF^* d_s + kF^* d_i + \eta d_v + \delta I^* d_v, \\
A_3 &= d_i d_v d_s + \delta I^* d_i d_v + \eta d_i d_v + kF^* d_i d_s + \delta I^* d_i kF^* \\
&\quad + kF^* \eta d_i - \beta \widehat{T}_s \mu(\eta + \delta I^*).
\end{aligned}
$$

When $A_3 > 0$ together with $A_1, A_2 > 0$, roots of the characteristic equation are negative.

$A_3 > 0$ implies $\frac{\lambda \beta \mu(\eta + \delta I^*)}{d_i(d_v + kF^*)(d_s + \eta + \delta I^*)(d_s - \alpha I^*)} < 1$ that is $R_c < 1$, where

$$R_c = \frac{\lambda \beta \mu(\eta + \delta I^*)}{d_i(d_v + kF^*)(d_s + \eta + \delta I^*)(d_s - \alpha I^*)} \qquad (7.23)$$

is the basic reproduction number in the presence of drugs.

Theorem 7.2.1 *The disease-free periodic orbit E_1 always exists and is stable whenever $R_c < 1$, but endemic periodic orbit E^* does not exist. When $R_c > 1$, E_1 becomes unstable and E^* does exist.*

7.2.3   Dynamics of the Drug

Suppose the drugs are given at fixed intervals. Let $\tau \equiv t_{k+1} - t_k$ be the period of the fusion inhibitor and $\sigma \equiv s_{k+1} - s_k$ be the period of the IL-2 (for $k \geq 1$).

We try to find out the general solution of the dynamics of drug

$$\frac{dF}{dt} = -m_F F, \quad t \neq t_k, \quad \Delta F = F^i, \quad t = t_k,$$
$$F(t_k^+) = F(t_k^-) + \Delta f_k, \quad t = t_k. \tag{7.24}$$

Therefore, the solution for t satisfying $t_k < t \leq t_{k+1}$ is

$$F(t) = F(t_k^+)e^{-m_F(t-t_k)}, \quad \text{for } t_k < t \leq t_{k+1}, \tag{7.25}$$

where $F(t)$ represents the rate of decay of the dose and $F(t_k^+)$ is the value at which the drug starts to decay instantaneously after the drug is injected.

The drug concentrations before and after the drug is taken are $F(t_k^+) = \frac{F^i(1-e^{-km_F\tau})}{1-e^{-m_F\tau}}$ and $F(t_{k+1}^-) = \frac{F^i(1-e^{-km_F\tau})e^{-m_F\tau}}{1-e^{-m_F\tau}}$. Thus for the limiting case the drug concentrations before and after one dosage are as follows: $\lim_{k\to\infty} F(t_k^+) = \frac{F^i}{1-e^{-m_F\tau}}$, $\lim_{k\to\infty} F(t_{k+1}^-) = \frac{F^i e^{-m_F\tau}}{1-e^{-m_F\tau}}$ and

$$F(t_{k+1}^+) = \frac{F^i e^{-m_F\tau}}{1-e^{-m_F\tau}} + F^i = \frac{F^i}{1-e^{-m_F\tau}}.$$

Here, we have found out that

$$F(t_k^+) - \frac{F^i}{1-e^{-m_F\tau}} = F^i\frac{1-e^{-km_F\tau}}{1-e^{-m_F\tau}} - \frac{F^i}{1-e^{-m_F\tau}} = -F^i\frac{e^{-km_F\tau}}{1-e^{-m_F\tau}} < 0.$$

We can conclude that the positive impulsive orbit for T-20 starts at $\frac{F^i e^{-m_F\tau}}{1-e^{-m_F\tau}}$ and ends at $\frac{F^i}{1-e^{-m_F\tau}}$.

As the same way, we can obtain for IL-2 that the positive impulsive orbit starts at $\frac{I^i e^{-m_I\sigma}}{1-e^{-m_I\sigma}}$ and ends at $\frac{I^i}{1-e^{-m_I\sigma}}$.

From the above observation, we can say that impulsive periodic orbits F^* and I^* for T-20 and IL-2 respectively satisfy

$$\frac{F^i e^{-m_F\tau}}{1-e^{-m_F\tau}} \leq F^* \leq \frac{F^i}{1-e^{-m_F\tau}}, \text{ and } \frac{I^i e^{-m_I\sigma}}{1-e^{-m_I\sigma}} \leq I^* \leq \frac{I^i}{1-e^{-m_I\sigma}}. \tag{7.26}$$

For simplification, we assume that σ is fixed.

Now, we find out the relation between drug dosage and dosing interval of T-20 therapy. We know that basic reproduction number $R_c < 1$ for disease-free steady state. Therefore, from (7.23),

$$\frac{\lambda\beta\mu(\eta+\delta I^*)}{d_i(d_v+kF^*)(d_s+\eta+\delta I^*)(d_s-\alpha I^*)} < 1,$$

we get

$$F^* > \frac{\lambda \beta \mu (\eta + \delta I^*) - d_i d_v (d_s + \eta + \delta I^*)(d_s - \alpha I^*)}{k d_i (d_s + \eta + \delta I^*)(d_s - \alpha I^*)}. \tag{7.27}$$

Theorem 7.2.2 *Let* $F_1 = \frac{X}{\phi}$ *and* $F_2 = \frac{X}{\psi}$, *where* $\phi = \frac{e^{-m_F \tau}}{1 - e^{-m_F \tau}}$ *and* $\psi = \frac{1}{1 - e^{-m_F \tau}}$, *then there exists the following two cases: (i) If* $F^i > F_1$, *the disease-free periodic orbit* (E_1) *exists and the endemic periodic orbit* (E^*) *does not exist for* $0 \le \tau < \tau_1$, *(ii) When the dosage satisfies* $0 \le F^i < F_2$, E_1 *is unstable and* E^* *exists for* $\tau > \tau_2$.

Proof From (7.26), we have $F^i \phi \le F^* \le F^i \psi$. If $F^i > F_1$, then $F^i > \frac{X(1 - e^{-m_F \tau})}{e^{-m_F \tau}}$
$\Rightarrow F^i \frac{e^{-m_F \tau}}{1 - e^{-m_F \tau}} > X \Rightarrow X < F^*$. Thus $X = \frac{\lambda \beta \mu (\eta + \delta I^*) - d_i d_v (d_s + \eta + \delta I^*)(d_s - \alpha I^*)}{k d_i (d_s + \eta + \delta I^*)(d_s - \alpha I^*)} < F^*$,
which implies $R_c < 1$ for $0 \le \tau < \tau_1$.

Therefore, the disease-free periodic orbit (E_1) exists.

Therefore for $\tau = \tau_1$, $\frac{e^{-m_F \tau_1}}{1 - e^{-m_F \tau_1}} F^i = X, \Rightarrow \tau_1 = \frac{1}{m_F} \log(1 + \frac{F^i}{X})$.

Again if $F^i < F_2$, then $F^i < X(1 - e^{-m_F \tau}) \Rightarrow \frac{F^i}{1 - e^{-m_F \tau}} < X \Rightarrow F^* < X$, which implies $R_c > 1$ for $\tau > \tau_2$.

Thus E_1 is unstable and E^* exists. Hence, for $\tau = \tau_2$, $\frac{F^i}{1 - e^{-m_F \tau_2}} = X \Rightarrow \tau_2 = -\frac{1}{m_F} \log(1 - \frac{F^i}{X})$.

Therefore, if drugs are taken repeatedly in the following pattern that the dosing interval τ and dosage F^i satisfy the above relation, then the disease can be controlled.

7.2.4 Numerical Simulation

In this section, we will perform various numerical simulations to establish and improve our analytic outcome. Parameter values are given in Table 7.2, which are chosen from Refs. [2, 5–7].

Here, we define the therapy strategy of antiviral drugs with perfect adherence (where the drugs are given at fixed intervals). The values of the parameter are taken from Table 7.2 and we choose the initial conditions of the variables as $T_s(0) = 180$, $T_l(0) = 10$, $T_i(0) = 10$, $V(0) = 1000$, and the unit of the concentration is mm^{-3}. Initially, we consider that the dosing interval of drug is fixed at $\tau = 0.5$, which means the drugs are taken two times a day. Under the chosen values of the parameters, basic reproduction number in the absence of drug is $R_0 = 2.6709 > 1$, which implies that disease-free equilibrium is unstable and disease persists. From Theorem 7.2.2,

Table 7.2 Parameters used in the model(5.43)

Parameter	λ	d_s	β	η	d_i	μ	d_v	k	α	δ
Default value	10	0.03	0.003	0.1	0.24	20	2.4	1	0.001	0.01

All parameter values are taken from [8–10]

we obtain two threshold values $F_1 = 3.0768$ and $F_2 = 1.8662$. Figure 7.3 shows the region of stability in the presence of drug therapy. If the drug dosage F^i is sufficiently large and dosing interval τ is sufficiently small, then E_1 is stable and E^* does not exist. However, if F^i is sufficiently small and τ is suitably large, then E_1 becomes unstable and E^* exists. However, the precise threshold lies between these two regions. Figure 7.4 shows the contour plot of basic reproduction number as a function of β and λ in the absence and presence of drug therapy respectively. Left panel demonstrates the change of R_0 as β and λ vary. Clearly, if $\lambda > 2.265$ and $\beta_1 < 0.000112$, then R_0 can be less than 1, and where disease-free equilibrium will be stable. When β increases and λ decreases, disease-free equilibrium losses its stability and moves towards unstable region. Right panel shows that in the presence of drug, disease-free equilibrium will be stable in spite of quite higher level of infection rate. If $\lambda > 6.083$ and $\beta_1 < 0.001224$, basic reproduction number in the presence of drug (R_c) can be less than 1, but whenever β increases and λ decreases, E_1 losses its stability and becomes unstable. From both contour plots, we can easily reach to the conclusion that drug enhances the stability of the system. Figure 7.5 shows the fluctuation of concentration for each model variable with respect to time in the presence of perfect adherence. Comparing with the figure in the absence of therapy (Inset Figure), we can easily say that combine therapy is actually effective to control

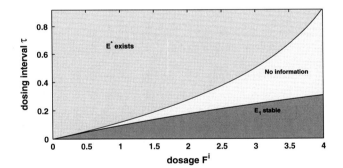

Fig. 7.3   Dose effect curve showing the region of stability

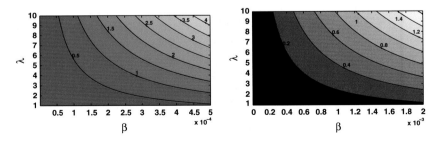

Fig. 7.4   *Left Panel* Contour plot of R_0 as a function of λ and β. *Right Panel* Contour plot of R_c as a function of λ and β

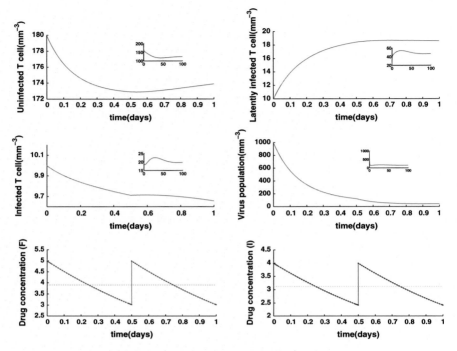

Fig. 7.5 System behavior for the perfect adherence with $F^i = 4$, $I^i = 3$ and $\tau = \sigma = 0.5$. *Inset* Trajectories of the system in the absence of therapy

the disease. In this case, we take $F^i = 4 (> F_1 = 3.0768)$ which is sufficiently large.

7.3 Effect of Chemokine Analog through Perfect Adherence

In this section, we have modified the explicit mathematical model as proposed by Smith [6], considering the perfect adherence behavior of chemokine analog in AIDS patients. Here, G_1 represents the concentration of viral Env subunit, $gp120$ in vivo, and C_{D4} denotes the concentration of $CD4$ receptor on T cell surface. Let C_1 be the concentration of the dimeric complex of $gp120$ and $CD4$ receptor, G_2 be the concentration of viral fusion protein, $gp41$, and C_{CR} be the concentration of the chemokine coreceptor on CD4$^+$ cells. Also, C_2 be the concentration of the combination of $gp41$ and $gp120$-$CD4$-chemokine coreceptor ternary complex, I denotes the concentration of infected CD4$^+$T cells, and V denotes the concentration of HIV virus. Here, R stands for concentration of chemokine analog. We have also examined the different possible outcomes at different drug doses. Thus our model becomes

$$\frac{dG_1}{dt} = bV - \mu_G G_1 - \beta_1 G_1 C_{D4},$$

$$\frac{dC_{D4}}{dt} = ps_T - \mu_T C_{D4} - \beta_1 G_1 C_{D4},$$

$$\frac{dC_1}{dt} = \beta_1 G_1 C_{D4} - d_1 C_1,$$

$$\frac{dG_2}{dt} = aC_1 - \mu_G G_2 - \beta_2 G_2 C_{CR},$$

$$\frac{dC_{CR}}{dt} = qs_T - \mu_T C_{CR} - \beta_2 G_2 C_{CR} - q_1 C_{CR} R + m_1 C'_{CR},$$

$$\frac{dC_2}{dt} = \beta_2 G_2 C_{CR} - d_2 C_2,$$

$$\frac{dI}{dt} = kC_2 - d_I I - \delta AI,$$

$$\frac{dV}{dt} = nd_I I - \mu_V V,$$

$$\frac{dA}{dt} = rV - d_A A,$$

$$\frac{dC'_{CR}}{dt} = q_1 C_{CR} R - \mu_T C'_{CR} - m_1 C'_{CR}, \tag{7.28}$$

for $t \neq t_j$. The impulsive effect is defined by,

$$\frac{dR}{dt} = -d_R R, \qquad t \neq t_j, \tag{7.29}$$

where the impulsive condition is

$$\triangle R = R_d, \qquad t = t_j. \tag{7.30}$$

In the system (7.28), b represents the multiplication capacity of $gp120$ in response to virus and a represents the successful exposure of $gp41$, since its exposure only after the attachment between $gp120$ and $CD4$ receptor is complete. The parameter μ denotes the decay of $gp120$ and $gp41$, and β_1 denotes the bonding force between $gp120$ and $CD4$ receptor. Furthermore, β_2 denotes the bonding force between dimeric complex of $gp120$ and $CD4$ receptor and also chemokine coreceptor. The source of susceptible CD4+T cells is represented by s_T, and p and q denote the number of $CD4$ receptors and chemokine coreceptors on one CD4+T cell respectively. The death rate of healthy CD4+T cells is μ_T and the death rate of infected CD4+T cells is d_I. The dissociation rates of C_1 and C_2 are d_1 and d_2 respectively, and for the sake of simplicity, it is assumed that after dissociation of C_1 and C_2, they will not return to their respective components. Here, n represents the number of virus particles that are produced by one infected CD4+ T cell. The clearance rate of free HIV virus is represented as μ_V. Antibody responses by A

are produced at a rate r and d_A indicates the death rate of antibody responses. The concentration of chemokine analog is represented by $R(t)$ in plasma. Parameter d_R represents the rate at which the drugs are cleared from intracellular compartments. The drug effects wear off at rate m_1. The drug dose R_d, that is taken at each impulse time t_j ($j = 1, 2, 3, \ldots$) is kept constant. Since drugs may be taken at either regular or irregular intervals, we have considered impulse time t_j to be fixed. The rate at which chemokine receptors on CD4$^+$T cells are blocked by chemokine analog is represented by C'_{CR}. Also, coreceptors are blocked at a rate q_1. The system (7.28) together with the systems (7.29) and (7.30) represents the dynamics with high drug concentration. Here, the dosing interval $\tau = t_{j+1} - t_j$ is fixed. Also, the degree of blocking of coreceptor by the chemokine analog depends on the drug concentration which has been reflected in the model. Here, we have only analyzed the models (7.28)–(7.30).

7.3.1 Analysis of the Model

In Absence of Drug: To study the model (7.28) together with (7.29) and (7.30), we first analyze the model in the absence of drug. When drug is not administered to the system, we have observed that there exists two steady states,

1. The disease-free state $\bar{E}$ and
2. The endemic state $\tilde{E}$.

7.3.1.1 The Disease-Free State

In the absence of drug, there may exist disease-free equilibrium $\bar{E}$ which is given by

$$(\bar{G}_1, \bar{C}_{D4}, \bar{C}_1, \bar{G}_2, \bar{C}_{CR}, \bar{C}_2, \bar{I}, \bar{V}, \bar{A}, \bar{C}'_{CR}) = (0, \frac{ps_T}{\mu_T}, 0, 0, \frac{qs_T}{\mu_T}, 0, 0, 0, 0, 0). \quad (7.31)$$

The characteristic equation for the disease-free equilibrium $\bar{E}$ is

$$(\Lambda + d_A)(\Lambda + \mu_T)^2(\Lambda + \mu_T + m_1)F(\Lambda) = 0, \quad (7.32)$$

where

$$F(\Lambda) = \begin{vmatrix} \Lambda + \mu_G + \beta_1\bar{C}_{D4} & 0 & 0 & 0 & 0 & -b \\ -\beta_1\bar{C}_{D4} & \Lambda + d_1 & 0 & 0 & 0 & 0 \\ 0 & -a & \Lambda + \mu_G + \beta_2\bar{C}_{CR} & 0 & 0 & 0 \\ 0 & 0 & -\beta_2\bar{C}_{CR} & \Lambda + d_2 & 0 & 0 \\ 0 & 0 & 0 & -k & \Lambda + d_I & 0 \\ 0 & 0 & 0 & 0 & -nd_I & \Lambda + \mu_V \end{vmatrix}.$$

From $F(\Lambda)$, we get

$$F(\Lambda) = \Lambda^6 + \eta_1\Lambda^5 + \eta_2\Lambda^4 + \eta_3\Lambda^3 + \eta_4\Lambda^2 + \eta_5\Lambda + \eta_6 = 0, \qquad (7.33)$$

where

$$
\begin{aligned}
\eta_1 &= d_1 + d_2 + d_I + 2\mu_G + \mu_V + \beta_1\bar{C}_{D4} + \beta_2\bar{C}_{CR} > 0, \\
\eta_2 &= (\mu_G + \beta_1\bar{C}_{D4})(\mu_G + \beta_2\bar{C}_{CR}) + (d_1 + d_2 + d_I + \mu_V) \times (2\mu_G + \beta_1\bar{C}_{D4} + \beta_2\bar{C}_{CR}) \\
&\quad + (d_1 + d_2)(\mu_V + d_I) + \mu_V d_I + d_1 d_2 > 0, \\
\eta_3 &= (\mu_G + \beta_1\bar{C}_{D4})(\mu_G + \beta_2\bar{C}_{CR})(d_1 + d_2 + d_I + \mu_V) \\
&\quad + (2\mu_G + \beta_1\bar{C}_{D4} + \beta_2\bar{C}_{CR})\{(d_1 + d_2)(\mu_V + d_I) \\
&\quad + (d_I\mu_V + d_1 d_2)\} + \{d_1 d_2(\mu_V + d_I) + d_I\mu_V(d_1 + d_2)\} > 0, \\
\eta_4 &= (2\mu_G + \beta_1\bar{C}_{D4} + \beta_2\bar{C}_{CR})\{d_1 d_2(\mu_V + d_I) + d_I\mu_V \\
&\quad \times (d_1 + d_2)\} + (\mu_G + \beta_1\bar{C}_{D4})(\mu_G + \beta_2\bar{C}_{CR}) \\
&\quad \times \{(d_1 + d_2)(\mu_V + d_I) + (d_I\mu_V + d_1 d_2)\} + d_1 d_2 d_I\mu_V > 0, \\
\eta_5 &= (\mu_G + \beta_1\bar{C}_{D4})(\mu_G + \beta_2\bar{C}_{CR})\{d_1 d_2(\mu_V + d_I) \\
&\quad + d_I\mu_V(d_1 + d_2)\} + d_1 d_2 d_I\mu_V(2\mu_G + \beta_1\bar{C}_{D4} + \beta_2\bar{C}_{CR}) > 0, \\
\eta_6 &= d_1 d_2 d_I\mu_V(\mu_G + \beta_1\bar{C}_{D4})(\mu_G + \beta_2\bar{C}_{CR}) - nkab\beta_1\beta_2 d_I\bar{C}_{D4}\bar{C}_{CR}. \qquad (7.34)
\end{aligned}
$$

For $\eta_6 > 0$, there exists no positive root and all roots have negative real part. Hence, the basic reproduction R_0 from (7.34) is

$$R_0 = \frac{nkab\beta_1\beta_2 pq s_T^2}{d_1 d_2 \mu_V(\mu_G\mu_T + ps_T\beta_1)(\mu_G\mu_T + qs_T\beta_2)}. \qquad (7.35)$$

Remark 7.3.1 The disease-free state $\bar{E}$ is locally stable if the basic reproduction number $R_0 < 1$ and the system is unstable when $R_0 > 1$.

The system E_0 is globally stable, if the system satisfies the following three conditions:

1. E_0 is locally asymptotically stable,
2. for $\frac{d\Gamma_1}{dt} = f(\Gamma_1, 0)$, Γ_1^0 is globally asymptotically stable, and
3. E_0 satisfies the Lyapunov–Lasalle theorem.

Now, we want to prove that the system is locally asymptotically stable. Thus we rewrite the system (7.28) in the form given below:

$$\frac{d\Gamma_1}{dt} = f(\Gamma_1, \Gamma_2), \quad \frac{d\Gamma_2}{dt} = g(\Gamma_1, \Gamma_2), \qquad (7.36)$$

where Γ_1 and Γ_2 are defined as follows:

$$\Gamma_1 = (C_{D4}, C_{CR}), \quad \Gamma_2 = (G_1, C_1, G_2, C_2, I, V, A, C'_{CR}). \quad (7.37)$$

Now, if we can show that $f(\Gamma_1, 0)$ is a limiting function of

$$\frac{d\Gamma_1}{dt} = f(\Gamma_1, \Gamma_2), \text{ i.e., } \lim_{t \to \infty} \Gamma_1 = \Gamma_1^*, \quad (7.38)$$

then we can easily conclude that the system is globally asymptotically stable. Furthermore,

$$C_{D4}(t) = \frac{ps_T}{\mu_T} - (\frac{ps_T}{\mu_T} - C_{D4}(0))e^{-\mu_T t} \longrightarrow \frac{ps_T}{\mu_T} = \bar{C}_{D4}(t), \text{ as } t \to \infty. \quad (7.39)$$

Also,

$$C_{CR}(t) \longrightarrow \frac{qs_T}{\mu_T} = \bar{C}_{CR}(t), \text{ as } t \to \infty. \quad (7.40)$$

Hence, we can conclude that the system is globally asymptotically stable. To prove the global stability of the disease-free system, we take the Lyapunov function in the form,

$$L = w_1 G_1 + w_2 C_1 + w_3 G_2 + w_4 C_2 + w_5 I + w_6 V + w_7 A + w_8 C'_{CR}, \quad (7.41)$$

where constants $w_i > 0$ for $i = 1, 2, \ldots, 8$. We can choose the constants as follows:

$$w_1 = \frac{\beta_1 ps_T a}{d_1(\beta_1 ps_T + \mu_G \mu_T)}, \quad w_2 = \frac{a}{d_1}, \quad w_3 = 1, \quad w_4 = \frac{\beta_2 ps_T + \mu_G \mu_T}{\beta_2 ps_T},$$

$$w_5 = \frac{n}{\mu_V}[\frac{\beta_1 ps_T ab}{d_1(\beta_1 ps_T + \mu_G \mu_T)} + w_7], \quad w_6 = \frac{1}{\mu_V}[\frac{\beta_1 ps_T ab}{d_1(\beta_1 ps_T + \mu_G \mu_T)} + w_7].$$

Hence, it can be written as

$$\frac{dL}{dt} \leq \Pi[R_0 + w'_7 - 1]C_2, \quad (7.42)$$

where $\Pi = \frac{d_2(\mu_T \mu_G + ps_T \beta_1)}{ps_T \beta_2}$ and $w'_7 = \frac{nkps_T \beta_1 w_7}{\mu_V d_2(\mu_T \mu_G + ps_T \beta_2)}$.

Now, $\frac{dL}{dt} \leq 0$ when $R_0 \leq (1 - w'_7)$, for $w_7 > 0$ which means in the presence of antibody responses, the disease-free equilibrium is stable. Also, $\frac{dL}{dt} = 0$ if $C_2 = 0$. Thus $I \to 0, V \to 0, G_1 \to 0, C_1 \to 0, G_2 \to 0$, when $t \to \infty$. Thus we can conclude that the disease-free state $\bar{E}$ is globally stable according to the Lyapunov–Lasalle theorem [6].

In the Presence of Drug: In the presence of drug, there exists no particular equilibrium point due to the administration of drug in impulse mode. However, we can calculate the equilibrium like orbit and these orbits are the impulsive analogs of equilibrium. There exists two equilibrium orbits (i) disease-free periodic orbit and

$$\bar{E}^* = (0, \bar{C}^*_{D4}, 0, 0, \bar{C}^*_{CR}, 0, 0, 0, 0, \bar{C}'^*_{CR}, R^*), \tag{7.43}$$

where

$$\bar{C}^*_{D4} = \frac{p_{ST}}{\mu_T}, \quad \bar{C}^*_{CR} = \frac{(\mu_T + m_1)q_{ST}}{\mu_T(\mu_T + m_1 + q_1 R^*)}, \quad \bar{C}'^*_{CR} = \frac{q q_1 s_T R^*}{\mu_T(\mu_T + m_1 + q_1 R^*)}. \tag{7.44}$$

Now, the disease-free periodic orbit $\bar{E}^*$ always exists if $R_0^d < 1$. The basic reproduction number in the presence of drug R_0^d is defined as,

$$R_0^d = \frac{nkab\beta_1\beta_2 pq s_T^2(\mu_T + m_1)}{\{d_1 d_2 \mu_V(\mu_T \mu_G + p_{ST}\beta_1)\}\{\mu_T \mu_G(\mu_T + m_1 + q_1 R^*) + q_{ST}\beta_2(\mu_T + m_1)\}}.$$

(ii) the endemic periodic orbit

$$\widetilde{E}^* = (\widetilde{G}^*_1, \widetilde{C}^*_{D4}, \widetilde{C}^*_1, \widetilde{G}^*_2, \widetilde{C}^*_{CR}, \widetilde{C}^*_2, \widetilde{I}^*, \widetilde{V}^*, \widetilde{A}^*, \widetilde{C}'^*_{CR}), \tag{7.45}$$

where

$$\widetilde{C}^*_{D4} = \frac{p_{ST}}{\mu_T + \beta_1 \widetilde{G}^*_1}, \quad \widetilde{C}^*_1 = \frac{p_{ST}\beta_1 \widetilde{G}^*_1}{d_1(\mu_T + \beta_1 \widetilde{G}^*_1)}, \quad \widetilde{A}^* = \frac{r\widetilde{V}^*}{d_A}, \quad \widetilde{I}^* = \frac{\mu_V \widetilde{V}^*}{nd_I},$$

$$\widetilde{C}'^*_{CR} = \frac{q_1 R^* \widetilde{C}^*_{CR}}{\mu_T + m_1}, \quad \widetilde{V}^* = \frac{(\mu_G \mu_T + p_{ST}\beta_1 + \mu_G \beta_1 \widetilde{G}^*_1)\widetilde{G}^*_1}{b(\mu_T + \beta_1 \widetilde{G}^*_1)},$$

$$\widetilde{G}^*_2 = \frac{(\xi_1 - \xi_2 \widetilde{G}^*_1 - \xi_3 \widetilde{G}^{*2}_1 - \xi_4 \widetilde{G}^{*3}_1)\widetilde{G}^*_1}{\xi(\mu_T + \beta_1 \widetilde{G}^*_1)}, \quad \widetilde{C}^*_2 = \frac{(\chi_1 + \chi_2 \widetilde{G}_1 + \chi_3 \widetilde{G}^2_1 + \chi_4 \widetilde{G}^3_1)\widetilde{G}_1}{\chi_5 + \chi_6 \widetilde{G}_1 + \chi_7 \widetilde{G}^2_1},$$

$$\widetilde{C}^*_{CR} = \frac{[nkq_{ST} d_I d_A - d_2 \mu_V \widetilde{V}(d_I d_A + r\delta \widetilde{V})](\mu_T + m_1)}{nk\mu_T d_I d_A(\mu_T + m_1 + q_1 R^*)},$$

$$\tag{7.46}$$

and $\widetilde{G}^*_1$ can be defined from

$$l_4 \chi_4^2 \widetilde{G}_1^7 + 2l_4 \chi_3 \chi_4 \widetilde{G}_1^6 + [l_4(2\chi_2\chi_4 + \chi_3^2) - \beta_1 \chi_4(l_2 + l_3\beta_1)]\widetilde{G}_1^5$$
$$- [\mu_T \chi_4(l_2 + 2l_3\beta_1) + \beta_1 \chi_3(l_2 + l_3\beta_1) - 2l_4(\chi_2\chi_3 + \chi_1\chi_4)]\widetilde{G}_1^4$$
$$- [\beta_1 \chi_2(l_2 + l_3\beta_1) + \mu_T \chi_3(l_2 + 2l_3\beta_1) + l_3\mu_T^2 \chi_4 - l_1\beta_1^3 - l_4(2\chi_1\chi_3 + \chi_2^2)]\widetilde{G}_1^3$$
$$- [\beta_1 \chi_1(l_2 + l_3\beta_1) + \mu_T \chi_2(l_2 + 2l_3\beta_1) + l_3\mu_T^2 \chi_3 - 2l_4 \chi_1 \chi_4 - 3l_1\mu_T \beta_1^2]\widetilde{G}_1^2$$
$$- [\mu_T \chi_1(l_2 + 2l_3\beta_1) + l_3\mu_T^2 \chi_2 - l_4 \chi_1^2 - 3\chi_1\mu_T^2\beta_1]\widetilde{G}_1$$
$$- \mu_T^2(l_3\chi_1 - l_1\mu_T) = 0,$$

$$\tag{7.47}$$

where

$$l_1 = n^2 k^2 ab^4 \beta_1 \beta_2 pqs_T^2 d_I^2 d_A^2, \quad l_2 = nkab^2 \beta_1 \beta_2 d_2 ps_T d_I d_A,$$
$$l_3 = nkb^2 d_I d_2 d_I d_A (\mu_G \mu_T' + qs_T \beta_2), \quad l_4 = d_I d_2^2 \beta_2,$$
$$\mu_T' = \frac{\mu_T (\mu_T + m_1 + q_1 R^*)}{\mu_T + m_1}, \tag{7.48}$$

and χ_i and ξ_i are defined in (7.25). The unique positive root of (7.47) exists if

$$\frac{nkab\beta_1\beta_2 pqs_T^2}{d_I d_2 \mu_V (\mu_G \mu_T + ps_T \beta_1)(\mu_G \mu_T' + qs_T \beta_2)} < 1. \tag{7.49}$$

However, for the endemic equilibrium, $R_0^d > 1$. Thus if $R_0^d > 1$ there exists two positive roots (multiple roots are permissible). Now, for the existence of $\widetilde{G}_2^*$, there exists $\widehat{G}_1^*$ such that $0 < \widetilde{G}_1^* < \widehat{G}_1^*$, where

$$\widehat{G}_1^* = \frac{nkabps_T \beta_1}{d_I d_2 \mu_V (\mu_G \mu_T + ps_T \beta_1)} > 1, \tag{7.50}$$

otherwise $\widehat{G}_1^* < 0$ and there exists no endemic equilibrium. Also, when $R_0^d > 1$, and for $\widehat{G}_1^* > 0$, there exists multiple positive roots and which is permissible. But the only positive root of $\widetilde{G}_1^*$ exists for $0 < \widetilde{G}_1^* < \widehat{G}_1^*$ which satisfies $\widetilde{G}_2^* > 0$. Hence, there exists only one endemic equilibrium.

7.3.2   Drug Dynamics

The impulsive equation

$$\frac{dR}{dt} = -d_R R, \quad t \neq t_j, \quad \Delta R = R_d, \quad t = t_j. \tag{7.51}$$

Here, $\Delta R_{j+1} \geq 0$ and R_d is the first dose of the drugs. The administration of drug in impulse mode is assumed to be fixed, which means that the drug is administered at regular dosing interval. Also, R_d is fixed, which represents the drug dosage with perfect adherence. The dosing interval can be considered as $\tau = t_{j+1}^- - t_j^+$ and thus in that interval $t \in [t_j^+, t_{j+1}^-]$, the solution of the system (7.51) is

$$R(t) = R(t_j^+)e^{-d_R(t - t_j)}, \quad t \in [t_j^+, t_{j+1}^-]. \tag{7.52}$$

For $t \in t_{j+1}^-$, $R(t_{j+1}^-) = R(t_j^+)e^{-d_R(t_{j+1} - t_j)} = R(t_j^+)e^{-d_R \tau}$. Also, $R(t_{j+1}^+) = R(t_{j+1}^-) + R_d$. Then the solution of the above equation along with the initial condition is $R(t_{j+1}^+) = R(t_j^+)e^{-d_R\tau} + R_d$. Now, for $j = 0$, $R(t_0) = R_d$. Thus we have derived the general solution of the maximal drug dose for the solution. It is clear that the solution does not depend on the time of dose to be given, but the interval and dosage are more important. Now the drug concentration

$$R(t) = (\sum_{i=0}^{j} r^{j-i} R_d)e^{-d_R(t-t_j)}, \quad \text{where } k = \lfloor \frac{t}{\tau} \rfloor$$

$$\text{and } t \in [t_j^+, t_{j+1}^-], \tag{7.53}$$

where $R(t)$ denotes the dynamics of the drug concentration in the interval $[t_j^+, t_{j+1}^-]$,

$$R(t) = R_d \frac{1 - r^{1+j}}{1 - r} e^{-d_R(t-t_j)}. \tag{7.54}$$

Now for a given dosing interval, drug concentration of the particular drug dose just before and after drug administration can be taken as follows:

$$\lim_{k\to\infty} R(t_k^+) = \frac{R_d}{1 - r}, \quad \text{and} \quad \lim_{k\to\infty} R(t_{k+1}^-) = \frac{R_d r}{1 - r}.$$

Thus the positive impulsive orbit is

$$\frac{R_d r}{1 - r} < R(t) < \frac{R_d}{1 - r}. \tag{7.55}$$

For the impulsive value of R, i.e., $R(t)$, we have

$$\tau > \frac{1}{d_R} \ln(\frac{R_d + R(t)}{R(t)}) = \tau_1. \tag{7.56}$$

Also,

$$\tau < \frac{1}{d_R} \ln(\frac{R(t)}{R(t) - R_d}) = \tau_2.$$

$$\frac{1}{d_R} \ln(\frac{R(t) + R_d}{R(t)}) < \tau < \frac{1}{d_R} \ln(\frac{R(t)}{R(t) - R_d})$$

$$\Rightarrow \tau_1 < \tau < \tau_2. \tag{7.57}$$

Remark 7.3.2 If we can restrict the dosing interval of τ satisfying the condition $0 \le \tau < \tau_1$ for fixed drug dosage, then the disease-free periodic orbit will be stable for the system. If $\tau > \tau_2$, the disease progression continues, even if drug is administered at fixed intervals. Thus, maintenance of optimum dosage regimen is essential in order to control the disease effectively. Furthermore, if the dosing interval

τ satisfies the condition $\tau_1 < \tau < \tau_2$, then endemic periodic orbit will be stable in the presence of drug dose and thus, disease progression can be controlled.

Now, in the presence of drug therapy, the basic reproductive ratio

$$R_0^d = \frac{nkab\beta_1\beta_2 pqs_T^2(\mu_T + m_1)}{\{d_1d_2\mu_V(\mu_T\mu_G + ps_T\beta_1)\}\{(\mu_T + m_1)(\mu_T\mu_G + qs_T\beta_2) + \mu_G\mu_Tq_1R(t)\}}. \tag{7.58}$$

Now we assume $\frac{e^{-d_R\tau}}{1-e^{-d_R\tau}} = \phi, \quad \frac{1}{1-e^{-d_R\tau}} = \psi$. Substituting this in (7.55), we get

$$R_d\phi < R(t) < R_d\psi. \tag{7.59}$$

Disease-free equilibrium $\bar{E}^*$ exists if $R_0^d < 1$,

$$\Rightarrow R_d > \frac{(\mu_G\mu_T + qs_T\beta_2)(\mu_T + m_1)}{\mu_G\mu_Tq_1\phi}(R_0 - 1) = R_1. \tag{7.60}$$

Also, for $R_0^d > 1$,

$$\Rightarrow R_d < \frac{(\mu_G\mu_T + qs_T\beta_2)(\mu_T + m_1)}{\mu_G\mu_Tq_1\psi}(R_0 - 1) = R_2. \tag{7.61}$$

Remark: From the above analytical findings, it should be concluded that the disease-free periodic orbit $\bar{E}^*$ is stable, if the drug dose satisfies the condition $R_d > R_1$, i.e., when the drug dose is sufficiently high. However, the infection persists and reaches an endemic state if drug dose satisfies the condition $R_d < R_2$. The above two conditions imply that if the drug dose satisfies the relation $R_2 < R(t) < R_1$, then the disease progression can be restricted with reduction in viral load in the affected patient.

7.3.3 Cell Count in Extreme Cases

In this section, we verify the effect of perfect adherence of the drug on the immune cells and on virions. Let, the total CD4$^+$T cell count be CD_{tot}. Then,

$$CD_{tot}(t) = C_{D4}(t) + C_{CR}(t) + C'_{CR}(t).$$
$$\frac{dCD_{tot}(t)}{dt} \leq (p + q)s_T - \mu_T[C_{D4}(t) + C_{CR}(t) + C'_{CR}(t)],$$
$$\Rightarrow C_{D4}(t) + C_{CR}(t) + C'_{CR}(t) \leq \frac{(p + q)s_T}{\mu_T}, \forall t \text{ (see Lemma 1)}. \tag{7.62}$$

$$CD_{tot}(t) \leq \frac{(p+q)s_T}{\mu_T}. \tag{7.63}$$

Hence, the total CD4$^+$T cell count will be less than $\frac{(p+q)s_T}{\mu_T}$ when infection is present.

Since the infected CD4$^+$T cells are produced from the uninfected CD4$^+$T cells, the maximum value of infected cell population will be $I(t) \leq \frac{(p+q)s_T}{\mu_T}$. Now, for the virions,

$$V(t) \leq \frac{nd_I(p+q)s_T}{\mu_T\mu_V}. \tag{7.64}$$

Thus the virions must satisfy the inequations (7.64).

Also, for the antibody response during the treatment,

$$A(t) \leq \frac{rnd_I(p+q)s_T}{\mu_T\mu_V d_A}. \tag{7.65}$$

Now, during the treatment,

$$\frac{dC_{CR}(t)}{dt} = qs_T - (\mu_T + q_1R^*)C_{CR} - \beta_2 G_2 C_{CR} + m_1 C'_{CR},$$

$$\Rightarrow C_{CR}(t) \leq \frac{qs_T + \frac{m_1(p+q)s_T}{\mu_T}}{\mu_T + q_1 R_d(t,\tau)} \quad [R(t) > R_d\phi = R_d(t,\tau)]$$

$$\longrightarrow 0, \tag{7.66}$$

as $t \to \infty$ when τ is fixed or $\tau \to 0$ for sufficiently small t.

For the concentration of viral Env subunit, gp120 in vivo,

$$G_1(t) \leq \frac{nbd_I(p+q)s_T}{\mu_T\mu_G\mu_V}. \tag{7.67}$$

For the concentration of the dimeric complex of gp120 and CD4 receptor,

$$C_1(t) \leq \frac{\beta_1 nbd_I(p+q)^2 s_T^2}{\mu_T^2\mu_V\mu_G d_1}. \tag{7.68}$$

For the concentration of viral fusion protein, gp41

$$G_2(t) \leq \frac{nabd_I\beta_1(p+q)^2 s_T^2}{\mu_T^2\mu_V\mu_G^2 d_1}. \tag{7.69}$$

Now during the treatment,

$$\frac{dC_{CR}}{dt} + \frac{dC'_{CR}}{dt} = qs_T - \mu_T(C_{CR} + C'_{CR}) - \beta_2 G_2 C_{CR}$$

$$\geq [qs_T - \frac{nabd_I\beta_1\beta_2(p+q)^2s_T^2}{\mu_V\mu_T^2\mu_G^2d_1}\frac{qs_T\mu_T + m_1(p+q)s_T}{\mu_T(\mu_T + q_1R_d(t,\tau))}] - \mu_T(C_{CR} + C'_{CR})$$

$$\Rightarrow C_{CR}(t) \to 0, \quad C'_{CR}(t) \to \frac{qs_T}{\mu_T}, \tag{7.70}$$

as $t \to \infty$ when τ is fixed or $\tau \to 0$ for sufficiently small t. Also,

$$\frac{dC_2}{dt} = \beta_2 G_2 C_{CR} - d_2 C_2 \leq \frac{nab\beta_1\beta_2 d_I(p+q)^2 s_T^3[q\mu_T + m_1(p+q)]}{\mu_T^3\mu_G^2\mu_V d_1[\mu_T + q_1 R_d(t,\tau)]} - d_2 C_2$$

$$\Rightarrow C_2(t) \leq \frac{nab\beta_1\beta_2 d_I(p+q)^2 s_T^3[q\mu_T + m_1(p+q)]}{\mu_T^3\mu_G^2\mu_V d_1 d_2[\mu_T + q_1 R_d(t,\tau)]}$$

$$\longrightarrow 0, \tag{7.71}$$

as $t \to \infty$ when τ is fixed or $\tau \to 0$ for sufficiently small t.

Remark: When the duration of treatment is sufficiently long (i.e., $t \to \infty$) with the drug being administered at fixed intervals, the percentage of chemokine coreceptors blocked by chemokine analog attains a maximum value suggesting that ternary complex formation with the involvement of viral protein is not possible ($C_2(t) \to 0$). The same condition also holds if the drug is given frequently (i.e., $\tau \to 0$) and the treatment is continued for a very short duration. Thus for effective management of the disease, the therapeutic intervention can adopt any one of the above-mentioned approaches. The ultimate outcome will be inhibition of viral entry into host cell and the host cell population remaining unaffected in the presence of viruses as a maximum number of chemokine receptors remain blocked ($C'_{CR}(t) \to \frac{qs_T}{\mu_T}$).

7.3.4 Numerical Simulation

In our numerical simulation, we have described the perfect drug adherence of chemokine analog. All the parameters are taken from Table 7.1. We have assumed the initial conditions as $G_1(0) = 2100$, $C_{D4}(0) = 800$, $C_1(0) = 0$, $G_2(0) = 1050$, $C_{CR}(0) = 800$, $C_2(0) = 0$, $I(0) = 0$, $V(0) = 50$, $A(0) = 100$, $C'_{CR}(0) = 0$ and the unit of the concentration is mm^{-3}.

Figure 7.6 (left and right panels) shows the contour graph of the basic reproduction number in the absence (R_0) and presence (R_0^d) of drug therapy as a function of β_2 and k respectively. Comparing both the plots, it appears that if the drug is introduced, the disease-free state will exist inspite of high infection rate k though the bonding force (β_2) is high. Figure 7.7 shows the region of stability and instability of the system with respect to the drug dosage (R_d) and the dosing interval (τ). In this figure, there are three regions. If the drug is administered at a dose $R_d > R_1$ and the dosing interval

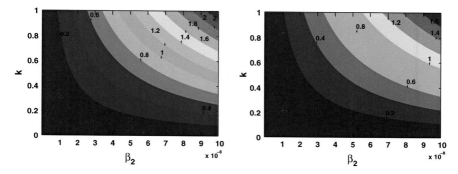

Fig. 7.6 *Left Panel* Contour plot of R_0 as a function of β_2 and k. *Right Panel* Contour plot of R_0^d as a function of β_2 and k

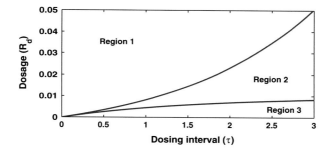

Fig. 7.7 Region of stability and instability. *Region 1* The disease-free state is stable if the drug dose is high and administered at moderate dosing interval. *Region 2* The endemic periodic orbit is stable and virus is controlled by the drug dose. *Region 3* The endemic periodic orbit exists if the drug dosage is low and the dosing interval is large, but its population cannot be controlled

$\tau < \tau_1$, the basic reproductive ratio may attain a value less than 1, when disease-free orbit or Region 1 is found to exist. However, if the drug is administered at a dose below the threshold value ($R_d < R_2$) and dosing interval $\tau > \tau_2$, the basic reproductive ratio remains above 1 and the disease progression continues to reach an endemic state (Region 3), where the virus cannot be controlled. From the above discussion, it can be suggested that if the drug dose can be maintained within R_1 and R_2 and dosing interval $\tau_1 < \tau < \tau_2$, the disease progression can be restricted which implies that viruses may be controlled (Region 2). Thus, the disease can be controlled effectively, when the drug is administered at fixed doses above threshold value ($R_2 < R_d < R_1$). and at regular intervals in an impulse mode. In Fig. 7.8, we observe that the system moves to its disease-free state, when the basic reproduction number R_0 becomes 0.722, for $k = 0.1$, but the system moves towards its endemic state, when k increases to its value 0.2, and thus basic reproduction number becomes 1.444. As a consequence, from this figure, we can conclude that the disease persists for increasing value of k. However, our foremost objective is to find out the result of impulsive drug effect to

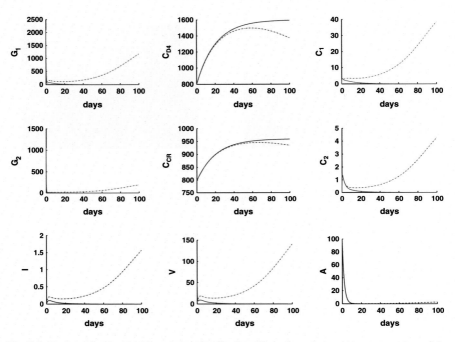

Fig. 7.8 Trajectories (marked by $-$) showing the time-dependent changes in concentration of the model variables when $R_0 < 1$ and trajectories (marked by $-.-$) showing the time-dependent change in concentration of the model variables when $R_0 > 1$. All parameters are same as in Table 7.3

Table 7.3 Parameters used in the models (7.28)–(7.30)

Parameter	b	β_1	μ_G	p	s_T	μ_T	d_1	a	β_2	q	
Default value	10	0.03	1500	0.02	0.002	0.24	0.002	0.2	0.02	0.0008	
Parameter	q_1	m_1	d_2	k	d_I	δ	n	μ_V	r	d_A	d_R
Default value	10	0.03	1500	0.02	0.002	0.24	0.002	0.2	0.02	0.0008	1

All parameter values are taken from [6, 11, 12]

the system when the endemic periodic orbit does exist. On that viewpoint, we have studied the system behavior in the presence of drug which is reflected in Fig. 7.9. If drug is administered at a fixed dose of $R_d = 0.6$ and $\tau = 1$, it is found from Fig. 7.9 that the complex formation between viral protein and chemokine coreceptor declines with time. The figure also reveals that the percentage of coreceptors blocked by the analog reaches a steady maximum value. Moreover, the percentage of the coreceptors that remains free (i.e., not blocked by the analog) attains a minimum value. This indicates that the viral population fails to infect new uninfected CD4$^+$T cell. Thus

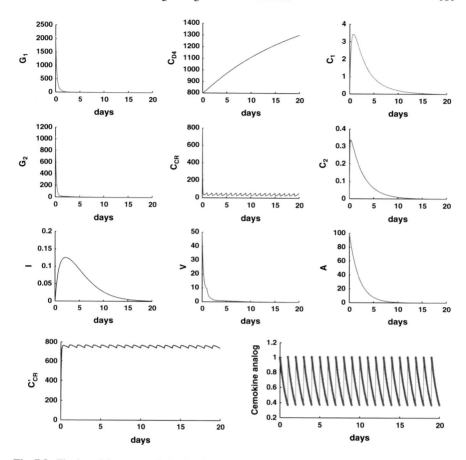

Fig. 7.9 The impulsive system behavior for the perfect adherence with $R_d = 0.6$ and $\tau = 1$

the infected cell population is reduced and the uninfected cell population approaches to its initial value. Since infection cannot be maintained at this dosage regimen of the chemokine analog, the antibody response also diminishes.

References

1. Smith, R.J.: Explicitly accounting for antiretroviral drug uptake in theoretical HIV models predicts long-term failure of protease-only therapy. J. Theor. Biol. **251**(2), 227–237 (2008)
2. Song, B., Lou, J., Wen, Q.: Modeling two different therapy strategies for drug T-20 on HIV-1 patients. Appl. Math. Mech. **32**(4), 419–436 (2011)
3. Bainov, D.D., Simeonov, P.S.: Systems with Impulsive Effect. Ellis Horwood Ltd, Chichester (1989)

4. Smith, R.J., Wahl, L.M.: Distinct effects of protease and reverse transcriptase inhibition in an immunological model of HIV-1 infection with impulsive drug effects. Bull. Math. Biol. **66**(5), 1259–1283 (2004)
5. Smith, R.J.: Adherence to antiretroviral HIV drugs: how many doses can you miss before resistance emerges? Proc. R. Soc. B **273**, 617–624 (2006)
6. Lou, J., Smith, R.J.: Modelling the effects of adherence to the HIV fusion inhibitor enfuvirtide. J. Theor. Biol. **268**, 1–13 (2011)
7. Perelson, A.S., Krischner, D.E., De-Boer, R.: Dynamics of HIV infection of CD4 T cells. Math. Biosc. **114**, 81–125 (1993)
8. Perelson, A.S., Neuman, A.U., Markowitz, M.: J.M., Leonard, Ho, D.D.: HIV 1 dynamics in vivo: viron clearance rate, infected cell life span, and viral generation time. Science **271**, 1582–1586 (1996)
9. Yang, J., Wang, X., Zhang, F.: A differential equation model of HIV infection of CD4$^+$ T cells with delay. Disc. Dyn. Nat. Soc. (2008). Article ID 903678, 16 pages doi:10.1155/2008/903678.179
10. Bonhoeffer, S., Coffin, J.M., Nowak, M.A.: Human immunodeficiency virus drug therapy and virus load. J. Virol. **71**, 3275–3278 (1997)
11. Smith, R.J., Wahl, L.M.: Drug resistance in an immunological model of HIV-1 infection with impulsive drug effects. The Bull. Math. Biol. **67**(4), 783–813 (2005)
12. Smith, R.J., Aggarwala, B.D.: Can the viral reservoir of latently infected CD4$^+$T cells be eradicated with antiretroviral HIV drugs? J. Math. Biol. **59**, 697–715 (2009)

Chapter 8
Mathematical Models in Stochastic Approach

Abstract Stochastic models in HIV are widely discussed in this chapter under different mathematical modelling perceptive . Here, we have studied various stochastic models of the deterministic system. We have estimated the probability of extinction theoretically followed by numerical simulations of the results. Extinction phenomenon will theoretically establish the status of the immune system in the absence of infection, where there has been spontaneous recovery without any therapeutic intervention. Our ultimate objective is to explore critical parameters, whose natures play a key role in determining the outcome of infection and in particular whether the HIV population will persist or become extinct. We have also focused on the insight of T cell proliferation in the expected time to extinction. Our analysis concludes that the expected time to extinction of the HIV infection from quasi-stationarity is an increasing function of the population size. In this way, proliferation of T cells through mitosis when HIV invades, has contributed significantly for expected time to extinction, which reveals proper justification by our analytical and numerical results.

Keywords Antigenic stimulation · Homeostasis · Kolmogorov's forward equation · Diffusion approximation · Parameter estimation · Quasi-stationary distribution · Transition states · Marginal distribution

8.1 Impact for Antigenic Stimulation on T Cell Homeostasis

In this section, we study the stochastic model of the system (2.1.5) studied earlier. Also probability of extinction has been determined theoretically followed by numerical simulations of the results. Extinction phenomenon will theoretically establish the status of the immune system in the absence of infection, where there has been spontaneous recovery without any therapeutic intervention. Our ultimate objective is to explore critical parameters, whose natures play a key role in determining the outcome of infection and in particular whether the HIV population will persist or become extinct.

© Springer Science+Business Media Singapore 2015
P.K. Roy, *Mathematical Models for Therapeutic Approaches to Control HIV Disease Transmission*, Industrial and Applied Mathematics,
DOI 10.1007/978-981-287-852-6_8

Table 8.1 Hypothesized transition rates for the stochastic version

Event	Transition	Transition rate
Birth of a target cell	$(m, n) \to (m + 1, n)$	$\lambda_1 = s + \eta_1 m$
Death of a target cell	$(m, n) \to (m - 1, n)$	$\mu_1 = dm$
Birth of an infected cell	$(m, n) \to (m - 1, n + 1)$	$\nu_1 = \beta' mn$
Death of an infected cell	$(m, n) \to (m, n - 1)$	$\mu_2 = \alpha dn + \frac{kn^2}{n+\theta}$

The Basic Assumptions

A rapid time scale for the free virus dynamics $V \approx \frac{pI}{c}$ is given by,

$$\frac{dT}{dt} = s - dT - \beta' TI + \eta_1 T,$$
$$\frac{dI}{dt} = \beta' TI - [\alpha d + k\frac{I}{I+\theta}]I, \tag{8.1}$$

where $\beta' = \frac{p\beta}{c}$. There are two state variables namely the number of target cells $T(t)$ and the number of infected cells $I(t)$ at time t. They jointly take values in the state space $S = \{(m, n) : m = 0, 1, 2, \ldots, n = 0, 1, 2, \ldots\}$. The joint distribution of $T(t)$ and $I(t)$ at time t is denoted as $p_{m,n}(t) = P\{T(t) = m, I(t) = n\}$. We use this notation even when m and/or n are negative or when $m > N$ with the convention that $p_{m,n}(t)$ is equal to zero.

The model is based on the following four basic events, i.e., birth of a target cell, death of a target cell, birth of an infected cell, and death of an infected cell. The transition rates of the model are shown in Table 8.1.

8.1.1 Formulation of the Kolmogorov's Forward Equation

We are assuming that in a time interval of infinitesimally small length Δt, the probability of exactly one birth (or one death) is birth rate (or death rate) $\times (\Delta t) + o(\Delta t)$, and that of more than one event (birth and/or death) in $o(\Delta t)$. We also consider the probability $p_{m,n}(t + \Delta t)$, where $\Delta t \downarrow 0$. The Kolmogorov forward equations for the model can be written as,

$$p_{m,n}(t + \Delta t) = \lambda_1(m - 1, n)p_{m-1,n}(t)\Delta t + \mu_1(m + 1, n)p_{m+1,n}(t)\Delta t$$
$$+ \nu_2(m, n - 1)p_{m,n-1}(t)\Delta t + \mu_2(m - 1, n + 1)p_{m-1,n+1}(t)\Delta t$$
$$+ p_{m,n}(t) - M(m, n)p_{m,n}(t)\Delta t + o(\Delta t), \tag{8.2}$$

where $M(m, n) = \lambda_1(m, n) + \mu_1(m, n) + \nu_1(m, n) + \mu_2(m, n)$.

Note that all proceedings consisting of more than one birth or more than one death are incorporated in the $o(\Delta t)$ expression.

Therefore,

$$\frac{p_{m,n}(t + \Delta t) + p_{m,n}(t)}{\Delta t} = \lambda_1(m - 1, n)p_{m-1,n}(t) + \mu_1(m + 1, n)p_{m+1,n}(t)$$
$$+ v_2(m + 1, n - 1)p_{m+1,n-1}(t) + \mu_2(m, n + 1)p_{m,n+1}(t)$$
$$- M(m, n)p_{m,n}(t) + \frac{o(\Delta t)}{\Delta t}.$$

Hence

$$p'_{m,n}(t) = \lim_{\Delta t \to 0} \frac{p_{m,n}(t + \Delta t) + p_{m,n}(t)}{\Delta t} = \lambda_1(m - 1, n)p_{m-1,n}(t) + \mu_1(m + 1, n)p_{m+1,n}(t)$$
$$+ v_2(m + 1, n - 1)p_{m+1,n-1}(t) + \mu_2(m, n + 1)p_{m,n+1}(t) - M(m, n)p_{m,n}(t).$$

8.1.2 Finding the Time to Extinction of Infected Cells

We have assumed that infected cells become extinct. So, there seems to be an eventual absorption of $I(t)$ at 0. Let τ be a random variable denoting the time to extinction of the infected cells, i.e., as long as we have $(t \leq \tau)$, infected cells will exist.

Therefore, $P(t \leq \tau) = P\{T(t) > 0\}$ and $P(t \leq \tau) = P\{I(t) = 0\} = p_{.0}(t)$,

where $p_{.0} = \sum_{m=0}^{N} p_{mn}(t)$.

Let F be the c.d.f. of τ and f be the p.d.f. of τ.

$$F(t) = P\{I(t) > 0\} = \sum_{m=0}^{N} P\{T(t) = m, I(t) = 0\} = \sum_{m=0}^{N} p_{m.0}(t)$$

and $f(t) = \sum_{m=0}^{N} p'_{m.0}(t)$.

Putting $n = 0$ in the expression for $p'_{mn}(t)$ we have,

$$\sum_{m=0}^{\infty} p'_{m.0}(t) = (d\alpha + \frac{k}{1 + \theta}) \sum_{m=0}^{N} p_{m.1}(t),$$

$$= (d\alpha + \frac{k}{1 + \theta}) \sum_{m=0}^{N} P(T(t) = m, I(t) = 1) \qquad (8.3)$$

and

$$f(t) = (d\alpha + \frac{k}{1+\theta}) \sum_{m=0}^{N} P(I(t) = 1) = (d\alpha + \frac{k}{1+\theta}) p_{.1},$$

where $p_{.1} = \sum_{m=0}^{N} p_{m,1}.$ \hfill (8.4)

To find the p.d.f. $f(t)$ and hence the expected time to extinction, we need to compute $p_{.1}(t)$.

8.1.3 The Distribution of the Time to Extinction

To verify the intuition of eventual absorption of I(t) at '0', the explicit evaluation of the distribution of $(T(t), I(t)) \forall t \geq 0$ seems necessary but is not possible. So we look characteristics of the process that will give more detailed information than the bare fact that eventual absorption in the class $(m, 0) : m = 1, 2, \ldots$ is certain.

The process $T(t), I(t), t \geq 0$ has a unique limiting conditional distribution $q_{m,n}$, where

$$q_{m,n} = \lim_{t \to \infty} P\{T(t) = m, I(t) = n \mid I(t) \neq 0\}. \tag{8.5}$$

Let

$$q_{m,n} = \lim_{t \to \infty} P\{T(t) = m, I(t) = n \mid I(t) \neq 0\},$$

$$= \frac{p_{m,n}}{1 - \sum_{m=0}^{\infty} p_{m,1}}, \quad m = 0, 1, 2, \ldots; n = 0, 1, 2, \ldots. \tag{8.6}$$

$$q'_{m,n}(t) = \frac{p'_{m,n}}{1 - \sum_{m=0}^{\infty} p_{m,1}} + (-1)\frac{-p'_{m,n}}{(1 - \sum_{m=0}^{\infty} p_{m,1})^2} p_{m,n},$$

$$= \frac{p_{m,n}}{1 - \sum_{m=0}^{\infty} p_{m,1}} + \frac{p_{m,n}}{(1 - \sum_{m=0}^{\infty} p_{m,1})^2}(d\alpha + \frac{k}{1+\theta}) p_{m,1}(t). \tag{8.7}$$

The distribution of the time to extinction will be especially needed, when initial distribution is assumed to be the quasi-stationary distribution, i.e., $p_{m,n}(0) = q_{m,n} \forall m, n$. Let τ_Q be the time to extinction of infected cell derived from quasi-stationarity. We can also write $q'_{m,n}(t)$ as,

$$q'_{m,n}(t) = \frac{p_{m,n}}{1 - \sum\limits_{m=0}^{\infty} p_{m,1}} + \frac{p_{m,n}}{1 - \sum\limits_{m=0}^{\infty} p_{m,1}} (d\alpha + \frac{k}{1+\theta}) \sum_{m=0}^{\infty} q_{.1}. \tag{8.8}$$

Putting $q'_{m,n}(t) = 0$ in (8.8), we get,

$$p'_{m,n} = (d\alpha + \frac{k}{1+\theta}) q_{.1} p_{m,n}(t). \tag{8.9}$$

Let $p_{m,n}(t) = u(>0)$, thus we have,

$$\frac{du}{dt} = -(d\alpha + \frac{k}{1+\theta}) q_{.1} u,$$

$$\text{or,} \int \frac{1}{u} du = -(d\alpha + \frac{k}{1+\theta}) q_{.1} \int dt,$$

$$\text{or,} p_{m,n}(t) = e^{(d\alpha + \frac{k}{1+\theta}) q_{.1} t} \text{constant}. \tag{8.10}$$

For $t = 0$, we have $p_{m,n}(0) = q_{m,n}$ $p_{m,n}(t) = q_{m,n} e^{(d\alpha + \frac{k}{1+\theta}) q_{.1} t}$ and $p_{.n}(t) = q_{.n} e^{(d\alpha + \frac{k}{1+\theta}) q_{.1} t}$.

Using above form of $p_{.1}(t)$ in Eq. (2.16) we have,

$$f(t) = p'_{.0}(t) = (d\alpha + \frac{k}{1+\theta}) p_{.1}(t),$$

$$= (d\alpha + \frac{k}{1+\theta}) q_{.1} e^{-(d\alpha + \frac{k}{1+\theta}) p_{.1}(t)}, t > 0. \tag{8.11}$$

Thus, the distribution of the time to extinction is especially simple, when the initial distribution is equal to the quasi-stationary distribution. To be explicit, let us denote the time to extinction from quasi-stationarity by τ_Q. τ_Q has an exponential distribution and the expected time to extinction of the infected cells is given by, (Table 8.2)

$$E(\tau_Q) = \frac{1}{(d\alpha + \frac{k}{1+\theta}) q_{.1}}. \tag{8.12}$$

Table 8.2 Possible changes in the two-population system (8.1) with the probabilities

Change	Probability
$\Delta x_1 = [1, 0]^T$	$p_1 = (s + T\eta_1)\Delta t$
$\Delta x_2 = [-1, 0]^T$	$p_2 = (dT + TI\beta')\Delta t$
$\Delta x_3 = [0, 1]^T$	$p_3 = TI\beta'\Delta t$
$\Delta x_4 = [0, -1]^T$	$p_4 = I[d\alpha + \frac{KI}{I+\theta}]\Delta t$

8.1.4 Diffusion Approximation

In order to approximate the quasi-stationary distribution, we consider the two-dimensional process given by,

$$\frac{dT}{dt} = s - dT - \beta'TI + \eta_1 T,$$
$$\frac{dI}{dt} = \beta'TI - [\alpha d + k\frac{I}{I+\theta}]I. \qquad (8.13)$$

Considering Eq. (8.1), we get the equilibrium points $E(\widehat{T}, \widehat{I})$.

$$\widehat{T} = \frac{1}{\beta'}(d\alpha + k\frac{\widehat{I}}{\widehat{I}+\theta}),$$
$$\widehat{I} = \frac{-\Omega_2 + \sqrt{(\Omega_2)^2 + 4\Omega_1\Omega_3}}{2\Omega_1}, \qquad (8.14)$$

where $\Omega_1 = \beta'(d\alpha + k)$, $\Omega_2 = d\alpha\theta\beta' - s\beta' + (d - \eta_1)(d\alpha + k)$, $\Omega_3 = d^2\alpha\theta - d\alpha\theta\eta_1 - s\theta\beta'$ and $\Theta = \frac{s\beta'}{d\alpha(d-\eta_1)}$.

When $\Theta < 1$, the system is stable and for $\Theta > 1$ the system is unstable.

It is now of interest to find the mean change $E(\Delta x)$ and the covariance matrix $E(\Delta(x)(\Delta(x)'))$ for the time interval Δt. We neglect the terms of order $(\Delta t)^2$, where superscript $\prime$ is used to denote transpose. In the time interval t to $(t + \Delta t)$, the changes in the state variables $\widehat{S}$ and $\widehat{I}$ are denoted by ΔT and ΔI.

$$E(\Delta x) = \begin{pmatrix} \Delta T \\ \Delta I \end{pmatrix} = \sum_{j=1}^{4} p_j \Delta x = p_1 \Delta x_1 + p_2 \Delta x_2 + p_3 \Delta x_3 + p_4 \Delta x_4,$$

$$= b(x)\Delta t + o(\Delta t),$$

where

$$b(x) = \begin{pmatrix} s - dT - \beta'TI + \eta_1 T \\ \beta'TI - [\alpha d + \frac{KI}{I+\theta}]I \end{pmatrix}.$$

We approximate $B(\widehat{x})$ by elevating it at deterministic critical point $\widehat{x} = [\widehat{T}, \widehat{I}]$.
Now,

$$B(\widehat{x}) = \begin{pmatrix} \eta_1 - d - \beta'\widehat{I} - \beta'\widehat{T} & -\beta'\widehat{T} \\ \beta'\widehat{I} & -\frac{K\widehat{I}\theta}{(\widehat{I}+\theta)^2} \end{pmatrix}.$$

Suppose the parameter α, γ, θ, s, β, η_1, d, k, p and c are such that we have a stable quasi-equilibrium point. For large N the process $N^{\frac{1}{2}}(x - \widehat{x})$ is approximated as a stable bivariate Ornstein–Uhlenbeck process with local drift matrix $B(\widehat{x})$ and local covariance matrix $D(\widehat{x})$; then the stationary distribution of the Ornstein–Uhlenbeck process is bivariate normal with mean 0 and variance $\Sigma = \begin{pmatrix} \sigma_1 & \sigma_2 \\ \sigma_2 & \sigma_3 \end{pmatrix}$

$$B(\widehat{x})\sum + \sum B'(\widehat{x}) = -D(\widehat{x}).$$

where $\sigma_1 = \frac{1}{a_1}[\beta'\widehat{T}\sigma_2 - s - \eta_1\widehat{T}]$, $\sigma_2 = \frac{a_1(\beta')^2(\widehat{T})^2 + a_2\beta'\widehat{I}(s+\eta_1\widehat{T})}{(a_1-a_2\widehat{I})(a_1 a_2 - (\beta')^2\widehat{T})}$, $\sigma_3 = \frac{1}{a_2}[\beta'\sigma_2 + \beta'\widehat{T}]$, $a_1 = \eta_1 - d - \beta'\widehat{I}$ and $a_2 = k\frac{\theta}{(\theta+\widehat{I})^2}$.

Thus, the marginal distribution of the pathogen population size in quasi-stationarity approximately achieves consistency with the fact that the approximating normal distribution is modified by truncation at $\frac{1}{2}$.

Hence we have the following approximation: $q_{.n} \approx \frac{1}{\sqrt{\frac{\sigma_3}{N}}} \frac{\phi(\frac{n-\widehat{I}}{\sqrt{\frac{\sigma_3}{N}}})}{\Phi(\frac{\widehat{I}-0.5}{\sqrt{\frac{\sigma_3}{N}}})}$, where Φ and ϕ are respectively the standard normal c.d.f. and the standard normal p.d.f.

8.1.4.1   The Expected Time to Extinction

We locate the expected time to extinction $(E(\tau_Q))$ from quasi-stationary distribution. It is given by,

$$E(\tau_Q) = \frac{1}{(d\alpha + \frac{k}{1+\theta})q_{.1}} = \frac{\sqrt{\frac{\sigma_3}{N}}}{(d\alpha + \frac{k}{1+\theta})} \frac{\frac{\Phi(\widehat{I}-0.5)}{\sqrt{\frac{\sigma_3}{N}}}}{\phi(\frac{1-\widehat{I}}{\sqrt{\frac{\sigma_3}{N}}})}.$$

We present the expected time to extinction as a function of population size N, which is a decreasing function when the population size is approximately 30, as depicted in Fig. 8.2.

8.1.5   Numerical Illustration

Figure 8.1 represents the normalized simulated marginal distribution profile of the infected host in quasi-stationary state, when N varies ($N = 100$–1000). It is observed that at $N = 100$, the distribution profile is truncated and skewed while with the increasing value of N. The curve displays higher kurtosis indicating a narrower distribution of the total strength of target and infected population. From Fig. 8.2, we get the expected time to extinction of the infected population from quasi-stationarity, when the total population varies from $N = 10$–50. Here, it is found that the overall expected time to extinction increases for increasing N except at the initial stages up to $N = 30$, where a reduction is observed. This can be explained by the fact that when the population size is small, the number of infected CD4$^+$T cells is also expected to be less quantity. The immunity provided by the effector CD8$^+$T cells can control the rapid growth of the infected cells resulting in lowering of the time to extinction of the disease. However, at larger values of N, the number of infected class is also expected to rise and in such a situation the immunity provided by the effector cells

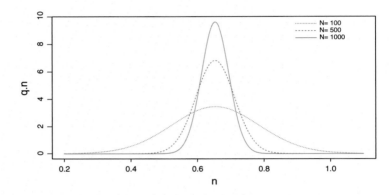

Fig. 8.1 Simulated marginal distribution of the infected host (I) in quasi-stationarity, together with approximation when $N = 100, 500, 1000$ and other parameters are $s = 100, d = 0.06, p = 70,$ $c = 1.3, \beta = 0.00015, \alpha = 2, k_0 = 0.9, a_E = 2.5, d_E = 1.5, \eta_1 = 0.01, \eta_2 = 0.03, \theta = 1$ and $\gamma = 0.03$

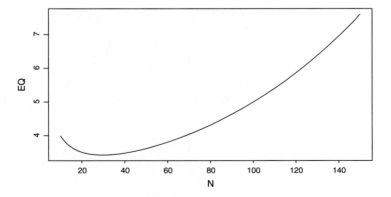

Fig. 8.2 Expected time to extinction of infected host (I) from quasi-stationarity as a function of N is plotted when total population $N = 10 - -150$ and other parameters are $s = 100, d = 0.06,$ $p = 70, c = 1.3, \beta = 0.00015, \alpha = 2, k_0 = 0.9, a_E = 2.5, d_E = 1.5, \eta_1 = 0.01, \eta_2 = 0.03,$ $\theta = 1$ and $\gamma = 0.03$

do not compensate for the spreading of the disease. The population of the infected cells rises and consequently the expected time to extinction of the disease increases.

8.1.6 Discussion

Introduction of stochasticity over the deterministic model also contributes to the novelty of the research under study. The numerical analysis reveals higher kurtosis with increasing total population size, which implies a higher contact rate of infection. The influence of the effector $CD8^+T$ cells has been considered here and considering the diffusion approximation and quasi-stationary distribution function, the expected time to extinction of the disease is estimated. The stochastic modeling thus provides a more accurate prediction in finding out the time to extinction and hence eradication of HIV disease pathogenesis.

8.2 Expected Time to Extinction of the Disease

8.2.1 Stochastic Version of the Model

Here we have introduced stochasticity in deterministic model of HIV considered by Zhou et al. [1]. We find the marginal distribution in quasi-stationary and the expected time to extinction, keeping in mind that the infected class has to be extinct. We have performed the numerical simulation to confirm our analysis.

Now we reduce the dimension by assuming a rapid time scale for the free virus dynamics, $V \approx qI/c$ [2, 3]. Thus the reduced ODE model can be written in the form of,

$$\frac{dT}{dt} = s - d_1 T + aT\left(1 - \frac{T}{T_{\max}}\right) - kTI + \rho I,$$

$$\frac{dI}{dt} = kTI - \delta I - \rho I, \tag{8.15}$$

where $k = \beta q/c$, all the parameters and variables are non negative.

Before describing the stochastic formulation, the logistic term can be written as, $A(T) - B(T) = aT(1 - \frac{T}{T_{\max}})$, where $A(T)$ is the birth rate and $B(T)$ is the death rate. Now, $A(T)$ and $B(T)$ can be written as $A(T) = a_1 T - a_2 T^2$ and $B(T) = b_1 T + b_2 T^2$ respectively, where a_i and b_i ($i = 1, 2$) are constants. So from the above relations we can state that $a = (a_1 - b_1)$ and $T_{\max} = \frac{a_1 - b_1}{a_2 + b_2}$.

Therefore, the above system of two equations can be written as,

$$\frac{dT}{dt} = s - d_1 T + (a_1 - b_1)T\left(1 - \frac{T}{\frac{a_1 - b_1}{a_2 + b_2}}\right) - kTI + \rho I,$$

$$\frac{dI}{dt} = kTI - \delta I - \rho I. \qquad\qquad (8.16)$$

8.2.2 The Stochastic Model Formulation

There are two state variables, namely the number of uninfected CD4$^+$T cells $T(t)$ and the number of infected such cells $I(t)$ at time t. They jointly take values in the state space $S = \{(m, n) : m = 0, 1, 2, \ldots; n = 0, 1, 2, \ldots\}$.

So, $p_{m,n}(t) = P\{T(t) = m, I(t) = n\}$ is the joint distribution of $T(t), I(t)$ at time t. If m and/or n are negative, $p_{m,n}(t)$ is then equal to 0.

The model is based on the following four basic events, i.e., production of CD4$^+$T cells primarily from bone marrow and secondarily antigen or mitogen stimulates T cells in presence of HIV virus, infection of CD4$^+$T cells, death of CD4$^+$T cells, and death of infected cells. Here we assume that the total population is N. Table 8.1 represents the hypothesized transition rates .

8.2.3 Description of the Transition States

The total number of population N is increased by unity. We can assume that in a infinitesimally small time interval Δt, the probability of the production of uninfected CD4$^+$T cells from precursors in the bone marrow and thymus is zero, but within the due time the constant rate of production of uninfected CD4$^+$T cells is s. But to make the population to be same, we should assume that there must be a natural death of uninfected CD4$^+$T cells. This phenomena is captured through the first two rows of the transition matrix. On the other hand, if there is an infection in the uninfected CD4$^+$T cells it can be balanced by an increase of infected CD4$^+$T cells. The infected class is generated by direct infection with replication of virus within the infected population. This occurs one at a time and so the increase in the infected class are reflected by the rise of unity in the transition state. If there is an infected CD4$^+$T cell, the natural death should be reflected through a natural birth of uninfected CD4$^+$T cells. At the end, naturally the recovery of infected host must reinstate to the uninfected CD4$^+$T cells.

8.2.3.1 Kolmogorov's Forward Equation

The Kolmogorov's forward equation for the model can be written as,

$$p_{m,n}(t + \Delta t) = p_{m-1,n}(t)\lambda_1(m - 1, n)\Delta t + p_{m+1,n}(t)\mu_1(m + 1, n)\Delta t$$
$$+ p_{m+1,n-1}(t)\lambda_2(m + 1, n - 1)\Delta t + p_{m,n+1}(t)\mu_2(m, n + 1)\Delta t$$
$$+ p_{m,n}(t)(1 - K(m, n)\Delta t) + o(\Delta t), \tag{8.17}$$

where $K(m, n) = \lambda_1(m, n) + \mu_1(m, n) + \lambda_2(m, n) + \mu_2(m, n)$. It is considered here that events consisting of more than one birth or more than one death are included in the $o(\Delta t)$ term. Thus,

$$\frac{p_{m,n}(t + \Delta t) - p_{m,n}(t)}{\Delta t} = p_{m-1,n}(t)\lambda_1(m - 1, n) + p_{m+1,n}(t)\mu_1(m + 1, n)$$
$$+ p_{m+1,n-1}(t)\lambda_2(m + 1, n - 1) + p_{m,n+1}(t)\mu_2(m, n + 1)$$
$$- p_{m,n}(t)K(m, n) + \frac{o(\Delta t)}{\Delta t}. \tag{8.18}$$

$$\text{Now, } p'_{m,n}(t) = \lim_{\Delta t \to 0} \frac{p_{m,n}(t + \Delta t) - p_{m,n}(t)}{\Delta t}. \tag{8.19}$$

Therefore, the Kolmogorov's forward equation for the model can be written as,

$$p'_{m,n}(t) = p_{m-1,n}(t)\lambda_1(m - 1, n) + p_{m+1,n}(t)\mu_1(m + 1, n)$$
$$+ p_{m+1,n-1}(t)\lambda_2(m + 1, n - 1) + p_{m,n+1}(t)\mu_2(m, n + 1)$$
$$- p_{m,n}(t)K(m, n). \tag{8.20}$$

8.2.3.2 Time to Extinction of Infected T Cells

The distribution of the time to extinction is an important measure in epidemiological perceptive, which can be determined from the solution of the Kolmogorov's forward equation (8.20).

Here, time to extinction of infected T cells be represented as τ is less than or equal to t; the number of infected individuals equals to zero, i.e., when we have $\{t \leq \tau\}$, infected T cells will exist.

So we can write $P(t \leq \tau) = P\{I(t) > 0\}$ and $P(\tau \leq t) = P\{I(t) = 0\} = p_{.0}(t)$, where $p_{.0}(t) = \sum_{m=0}^{N} p_{m,n}(t)$.

Thus the marginal probability of infected T cells equals to zero at time t that also equals to the cumulative distribution function of the time to extinction at time t.

Let the c.d.f. and p.d.f. of τ are denoted by F and f respectively.

Therefore, $F(t) = P\{I(t) = 0\} = \sum_{m=0}^{N} P\{T(t) = m, I(t) = 0\} = \sum_{m=0}^{N} p_{m,0}(t)$.

So, $f(t) = \sum_{m=0}^{N} p'_{m,0}(t) = p'_{.0}(t)$.

Putting $n = 0$ in Eq. (8.20) we have,

$$p'_{m,0}(t) = p_{m-1,0}(t)\lambda_1(m-1,0) + p_{m+1,0}(t)\mu_1(m+1,0)$$
$$+ p_{m+1,-1}(t)\lambda_2(m+1,-1) + p_{m,1}(t)\mu_2(m,1) - p_{m,0}(t)K(m,0),$$
$$= (s + (m-1)(a_1 - a_2(m-1)))p_{m-1,0}(t) + ((m+1)(b_1 + b_2(m+1)) + d_1))$$
$$+ p_{m+1,0}(t) + (\rho + \delta)p_{m,1}(t) - p_{m,0}(t)(s + m(a_1 - a_2m) + m((b_1 + b_2m) + d_1).$$

Taking summation on the both sides from $m = 0 \to \infty$ we get,

$$\sum_{m=0}^{\infty} p'_{m,0}(t) = s \sum_{m=0}^{\infty} p_{m-1,0}(t) + \sum_{m=0}^{\infty} (m-1)(a_1 - a_2(m-1))$$

$$= p_{m-1,0}(t) + \sum_{m=0}^{\infty} (m+1)(b_1 + b_2(m+1))p_{m+1,0} + d_1 \sum_{m=0}^{\infty} (m+1)p_{m+1,0}(t)$$

$$+ \delta \sum_{m=0}^{\infty} p_{m,1}(t) + \rho \sum_{m=0}^{\infty} p_{m,1}(t) - s \sum_{m=0}^{\infty} p_{m,0}(t)$$

$$- \sum_{m=0}^{\infty} m(a_1 - a_2m)p_{m,0}(t) - \sum_{m=0}^{\infty} m(b_1 + b_2m)p_{m,0}(t) - d_1 \sum_{m=0}^{\infty} mp_{m,0}(t),$$

$$= \delta \sum_{m=0}^{\infty} p_{m,1}(t) + \rho \sum_{m=0}^{\infty} p_{m,1}(t),$$

$$= (\delta + \rho) \sum_{m=0}^{\infty} P(T(t) = m, I(t) = 1), \tag{8.21}$$

$$\text{i.e., } p'_{.0}(t) = (\delta + \rho)p_{.1}(t). \tag{8.22}$$

So we can write,

$$f(t) = p'_{.0}(t) = (\delta + \rho)P\{I(t) = 1\} = (\delta + \rho)p_{.1}(t),$$

where $p_{.1}(t) = \sum_{m=0}^{\infty} p_{m,1}(t)$.
To find the expected time to extinction, we need to compute $p_{.1}(t)$.

8.2.3.3 The Distribution of the Time to Extinction

To find the absorption of $I(t)$ at 0, the distribution of $(T(t), I(t))$, $\forall t \geq 0$ is necessary but it is difficult. So, we look for the process that will give more detailed information regarding eventual absorption in the class $\{(m, 0) : m = 1, 2, \ldots\}$, which is assured in this perspective. Thus, we try to find out the quasi-limiting distribution. The process $\{T(t), I(t)\}, t \geq 0$ has a unique conditional distribution (conditioned being not absorbed) $q_{m,n}$,

where

$$q_{m,n}(t) = P\{T(t) = m, I(t) = n | I(t) \neq 0\} = \frac{p_{m,n}(t)}{1 - p_{.0}(t)},$$

$$m = 0, 1, 2, \ldots; n = 1, 2, \ldots$$

Now we derive a system of differential equation for $q_{m,n}(t)$. Differentiating the expression for $q_{m,n}(t)$ and using the relation (8.14) we obtain,

$$q'_{m,n}(t) = \frac{p'_{m,n}(t)}{1 - \sum_{m=0}^{\infty} p_{m,0}(t)} + (-1)\frac{-\sum_{m=0}^{\infty} p'_{m,0}(t)}{(1 - p_{.0}(t))^2} p_{m,n}(t),$$

$$= \frac{p'_{m,n}(t)}{1 - p_{.0}(t)} + \frac{p_{m,n}(t)}{(1 - p_{.0}(t))^2}(\delta + \rho)p_{.1}(t). \tag{8.23}$$

The distribution of the random variable τ depends upon the initial distribution of the infected T cell population. If at time t, the infection exists for a long time, then it can be assured that the distribution of the number of infected T cell population $I(t)$ can be described by the quasi-stationary distribution. On this stand point we are assuming that the initial distribution is quasi-stationary.

Let us consider τ_Q be the time to extinction of infected T cells. We can write $q'_{m,n}(t)$ as,

$$q'_{m,n}(t) = \frac{p'_{m,n}(t)}{1 - p_{.0}(t)} + (\delta + \rho)p_{.1}(t)\frac{p_{m,n}(t)}{(1 - p_{.0}(t))^2}. \tag{8.24}$$

Also we can write,

$$q'_{m,n}(t) = \frac{p'_{m,n}(t)}{1 - p_{.0}(t)} + (\delta + \rho)p_{.1}(t)\frac{q_{m,n}(t)}{(1 - p_{.0}(t))}. \tag{8.25}$$

Since the initial distribution is assumed to be quasi-stationary, i.e., $p_{m,n}(0) = q_{m,n}$, we can write,

$$q'_{m,n}(t) = \frac{p'_{m,n}(t)}{1 - p_{.0}(t)} + (\delta + \rho)q_{.1}(t)\frac{p_{m,n}(t)}{(1 - p_{.0}(t))}. \tag{8.26}$$

Putting $q'_{m,n}(t) = 0$, in Eq. (8.26) we get,

$$p'_{m,n}(t) = -(\delta + \rho)q_{.1}p_{m,n}(t), \tag{8.27}$$

i.e., $p_{m,n}(t) = q_{m,n}e^{-(\delta+\rho)q_{.1}t}$.
   So, $p_{.1}(t) = q_{.1}e^{-(\delta+\rho)q_{.1}t}$.
   Using the above form of $p_{.1}(t)$ in Eq. (8.14) we have,

$$f(t) = p'_{.0}(t) = ((\delta + \rho)q_{.1})e^{-((\delta+\rho)q_{.1})t}, t > 0.$$

Thus the expected time to extinction of the infected T cells has an exponential distribution and is equal to $E(\tau_Q) = \frac{1}{(\delta+\rho)q_{.1}}$.

Therefore, we can say that the expected time to extinction from quasi-stationary distribution is inversely proportional to the probability $q_{.1}$. It is also proportional to the sum of the recovery rate ρ and the disappearance rate δ of the infected T cell population. From the above expression we say that for fixed parameter values ρ and δ, τ_Q is determinable and quasi-stationary probability $q_{.1}$, which is the marginal distribution for one infected individual at time t conditioned not being absorbed.

8.2.4 Diffusion Approximation

Let us consider the two-dimensional process (8.16),

$$\frac{dT}{dt} = s - d_1 T + T(a_1 - b_1)\left(1 - \frac{T}{\frac{a_1-b_1}{a_2+b_2}}\right) - kTI + \rho I,$$

$$\frac{dI}{dt} = kTI - \delta I - \rho I. \tag{8.28}$$

If the total population N is sufficiently large, the quasi-stationary distribution is approximated by a bivariate normal distribution. Here we consider $y_1 = T$, $y_2 = I$ and the process $y(t) = (y_1(t), y_2(t))$.

The critical point of the deterministic model (8.15) is given by $\widehat{y} = (\widehat{y}_1, \widehat{y}_2)$, where $\widehat{y}_1 = \frac{\delta+\rho}{k}$, and $\widehat{y}_2 = \frac{1}{\delta}[s + a\frac{\delta+\rho}{k}(1 - \frac{\delta+\rho}{kT_{max}}) - \frac{d(\delta+\rho)}{k}]$.

Now we want to find the mean change $E(\Delta y)$ and the covariance matrix $E\left(\Delta y(\Delta y)^T\right)$ for the time interval Δt (neglecting the terms of order $(\Delta t)^2$).

In the time interval t to $t + \Delta t$, there are four possibilities of population change Δy by neglecting multiple births and deaths in the time interval t to $t+\Delta t$, which are of order $(\Delta t)^2$. The possibilities of population change are given in Table 8.2 along with their corresponding probabilities.

Here, $E(\Delta y) = \sum_{j=1}^{4} p_j \Delta y_j$,

$$= \begin{pmatrix} s + a_1 y_1 - a_2 y_1^2 + \rho y_2 - b_1 y_1 - b_2 y_1^2 - d_1 y_1 - k y_1 y_2 \\ k y_1 y_2 - (\rho + \delta) y_2 \end{pmatrix} \Delta t + o(\Delta t),$$

$$= b(y)\Delta t + o(\Delta t).$$

The Jacobian matrix of the vector $b(y)$ with respect to y is denoted by $B(y)$ and it is defined by,

$$B(y) = \frac{\partial b(y)}{\partial y} = \begin{pmatrix} (a_1 - b_1) - 2y_1(a_2 + b_2) - d_1 - ky_2 & \rho - ky_1 \\ ky_2 & ky_1 - \delta - \rho \end{pmatrix}.$$

The approximated value of $B(y)$ at the critical point $\widehat{y} = (\widehat{y}_1, \widehat{y}_2)$ is $B(\widehat{y}) = \begin{pmatrix} B_1 & B_2 \\ B_3 & B_4 \end{pmatrix}$,

where $B_1 = \frac{-s - \rho \widehat{y}_2}{\widehat{y}_1} - \frac{\widehat{y}_1 a}{T_{\max}}$, $B_2 = -\delta$, $B_3 = k\widehat{y}_2$, $B_4 = 0$.

Now,

$$E\left(\Delta y(\Delta y)^T\right) = \sum_{j=1}^{4} p_j \Delta y_j (\Delta y_j)^T = S(y),$$

$$= \begin{pmatrix} S_1 & S_2 \\ S_2 & S_3 \end{pmatrix},$$

where $S_1 = s + y_1(a_1 - a_2 y_1) + \rho y_2 + y_1(b_1 + b_2 y_1 + d_1) + ky_1 y_2$, $S_2 = -ky_1 y_2$, and $S_3 = 2(\delta + \rho)y_2$.

At the critical point $\widehat{y}$, the value of $S(y)$ is given by,

$$S(\widehat{y}) = \begin{pmatrix} 2(s + \widehat{y}_1(a_1 - a_2 \widehat{y}_1) + \rho \widehat{y}_2) & -k\widehat{y}_1 \widehat{y}_2 \\ -k\widehat{y}_1 \widehat{y}_2 & 2(\delta + \rho)\widehat{y}_2 \end{pmatrix}.$$

For large N, the process $N^{1/2}\{y(t) - \widehat{y}\}$ is approximated by a bivariate Ornstein–Uhlenbeck process with local drift matrix $B(\widehat{y})$ and local covariance matrix $S(\widehat{y})$. Thus its stationary distribution approximates the quasi-stationary distribution. The process $N^{1/2}\{y(t) - \widehat{y}\}$ is approximately bivariate normal with mean 0 and covariance matrix C, where the matrix C can be determined with the help of the local drift matrix $B(\widehat{y})$ and local covariance matrix $S(\widehat{y})$ through the relationship $B(\widehat{y})C + CB^T(\widehat{y}) = -S(\widehat{y})$, where B^T is the transpose of B. After solving the above equation we get,

$$C = \begin{pmatrix} C_1 & C_2 \\ C_2 & C_3 \end{pmatrix}.$$

The above result leads to the conclusion that in quasi-stationarity, the marginal distribution of the number of uninfected T cells and infected T cells are approximately $N(\mu_{y_1}, \sigma_{y_1})$ and $N(\mu_{y_2}, \sigma_{y_2})$ respectively, where

$$\mu_{y_1} = \widehat{y}_1, \sigma_{y_1} = \sqrt{(C_1/N)} \text{ and}$$
$$\mu_{y_2} = \widehat{y}_2, \sigma_{y_2} = \sqrt{(C_3/N)}.$$

Further, the covariance is approximated by $\sigma_{y_1 y_2} = C_2 = \frac{-(\delta + \rho)}{k}$,

$$C_1 = \frac{A}{B}, C_3 = \frac{B_1 C_2 + B_3 C_1 - k\widehat{y_1}\widehat{y_2}}{\delta}, \text{ where}$$

$A = s + \widehat{y_1}(a_1 - a_2\widehat{y_1}) + \rho\widehat{y_2} + \frac{\delta(\delta+\rho)}{k}, \ B = \frac{s+\rho\widehat{y_2}}{\widehat{y_1}} + \widehat{y_1}(a_2 + b_2).$
Note that the parameters should also satisfy $C_1 > 0, C_3 > 0$.

8.2.5 Expected Time to Extinction

In the deterministic version of the model, it is not possible to find out expected time to extinction of the disease though by this approach, recovery of the disease can be achieved by the immune system. Here, it is critical to mention that recovery of the infected cells and extinction of the disease, both are equally important. Stochastic approach is the only method by which expected time to extinction of the disease can be imagined, which is obviously related to the total number of cell populations.

Theorem 8.2.1 *The expected time to extinction of the infected cell is approximately*

$$E\left(\tau_Q\right) = \frac{1}{(\delta+\rho)q_{.1}} = \frac{\sigma_{y_2}}{(\delta+\rho)} \frac{\phi\left(\frac{\mu_{y_2}-1/2}{\sigma_{y_2}}\right)}{\varphi\left(\frac{1-\mu_{y_2}}{\sigma_{y_2}}\right)}$$

Proof The time to extinction (τ_Q) in quasi-stationarity follows an exponential distribution with parameter $(\delta+\rho)q_{.1}$, where $q_{.1}$ is the marginal quasi-stationary distribution of the number of infected cells, conditioned not being absorbed (i.e., $y_2 > 0$).

Using the diffusion approximation, the marginal distribution of the infected cells in quasi-stationarity will be univariate normal with mean μ_{y_2} and standard deviation σ_{y_2}.

To achieve consistency with the fact that $y_2 > 0$, the approximating normal distribution is modified by truncation at 0.5.

Hence we have the following approximation of $q_{.n}$,

$$q_{.n} \approx \frac{1}{\phi(\infty) - \phi\left(\frac{1/2-\mu_{y_2}}{\sigma_{y_2}}\right)} \frac{1}{\sqrt{2\pi}} \frac{1}{\sigma_{y_2}} e^{-\frac{(n-\mu_{y_2})^2}{2\sigma_{y_2}^2}}, \ (1/2 < n < \infty) \quad (8.29)$$

$$, = \frac{1}{1 - \phi\left(\frac{0.5-\mu_{y_2}}{\sigma_{y_2}}\right)} \frac{1}{\sigma_{y_2}} \varphi\left(\frac{n-\mu_{y_2}}{\sigma_{y_2}}\right), \quad (8.30)$$

i.e.,

$$q_{.n} \approx \frac{1}{\sigma_{y_2}} \frac{\varphi\left(\frac{n-\mu_{y_2}}{\sigma_{y_2}}\right)}{\phi\left(\frac{\mu_{y_2}-0.5}{\sigma_{y_2}}\right)}, \quad (8.31)$$

where ϕ and φ are the standard normal c.d.f. and the standard normal p.d.f. respectively.

Using approximation of (8.31), the expected time to extinction can be written as,

$$E\left(\tau_Q\right) = \frac{1}{(\delta+\rho)q_{.1}} = \frac{\sigma_{y_2}}{(\delta+\rho)} \frac{\phi\left(\frac{\mu_{y_2}-1/2}{\sigma_{y_2}}\right)}{\varphi\left(\frac{1-\mu_{y_2}}{\sigma_{y_2}}\right)}. \tag{8.32}$$

8.2.6 Numerical Simulation

For the numerical simulation of the marginal distribution in quasi-stationarity, we use Monte Carlo simulations techniques. For this purpose, we consider different population size (N) and different recovery rate of the infected CD4$^+$T cells. In the analytical derivation as well as in numerical simulation, we consider $a = (a_1 - b_1)$ and $T_{max} = \frac{a_1-b_1}{a_2+b_2}$, where a_i and b_i ($i = 1, 2$) are constants.

Figure 8.3 shows the marginal distribution in quasi-stationarity for the number of infected CD4$^+$T cells, which is truncated and positively skewed, i.e., non-normal at $N = 1500$. Figures 8.4 and 8.5 represent the marginal distribution of infected CD4$^+$T cells for different values of the total population N and recovery rate ρ. From Fig. 8.4, we can say that with increasing population size N, the number of infected CD4$^+$T cells also increases and the density of the infected CD4$^+$T cells is also enhanced. Figure 8.5 reflects the distribution of infected individuals for different values of ρ which shows that when we increase the recovery rate of the infected individual, the number of infected cell also decreases. So the skewness of the population decreases, where as the kurtosis gets higher value.

Figure 8.6 represents the expected time to extinction from marginal distribution of the infected population in quasi-stationarity for various population size. Here we

Fig. 8.3 The marginal distribution in quasi-stationarity of the number of infected CD4$^+$T cells, which is truncated and positively skewed, i.e., non-normal for $N = 1500$ of Eq. (8.16)

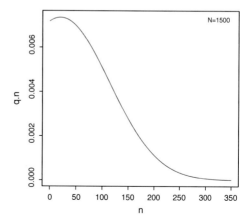

Fig. 8.4 The marginal
distribution in
quasi-stationarity of the
number of infected CD4$^+$T
cells for different values of N
of Eq. (8.16)

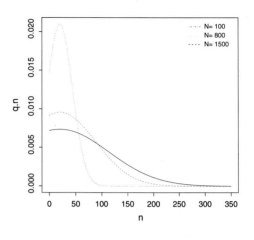

Fig. 8.5 The marginal
distribution in
quasi-stationarity of the
number of infected CD4$^+$T
cells for different values of ρ
of Eq. (8.16)

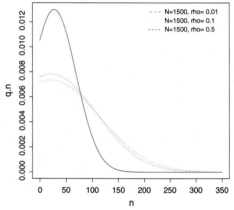

Fig. 8.6 The expected time
to extinction $E(\tau_Q)$ in
transition region is shown as
a function of N of Eq. (8.16),
all other parameters are same
as in Fig. 8.1 and Fig. 8.2

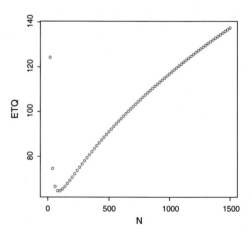

observe that when the population size N is very small, the expected time to extinction is very low but when N grows, the expected time to extinction increases rapidly. This type of phenomenon occurs because when the total population N is very small, it is expected that the number of infected cells is also very small. Furthermore, some of the infected cells become recovered, so the expected time to extinction becomes very low, but when the total population N is high, then the number of infected cells grows rapidly, which leads to the fact that the expected time to extinction is very high. Lastly, our numerical analysis shows that the dynamics of disease progression can be reduced by applying control measures in the drug adherence with increasing recovery rate of the infected $CD4^+T$ cells. With this, it has also been concluded that expected time to extinction of the disease is a function of the total number of population size.

8.2.7 Discussion

In this section, we have enriched the mathematical model of Zhou et al. [1] by introducing optimal control therapeutic approach for successful recovery of the disease. Our analytical and numerical results reveal that number of infected $CD4^+T$ cells decreases remarkably using systematic drug dose. By this, we get a better recovery of the infected class as it moves towards healthy class within a certain period of time. But for the estimation of expected time to extinction of the disease, it is not possible to analyze through deterministic version. For that purpose, we have studied the stochastic counterpart of the existing deterministic model of HIV. We have also derived the quasi-stationary distribution of infected individuals as well as expected time to extinction of the disease. In this regard, our research findings may add a new dimension in epidemiological study through estimated approximation of marginal density of the infected individuals and its time to extinction. In conclusion, it can be stated that stochastic modeling provides a more accurate prediction in finding out the expected time to extinction of infected population for the disease HIV/AIDS.

8.3   Insight of T Cell Proliferation in the Expected Time to Extinction

The expected time to extinction of the disease HIV/AIDS is a serious concern today. This expected time to extinction has a significant correlation with the increasing function of population size of $CD4^+T$ cells. When HIV invades in to the body, these healthy cells are stimulated by antigen or mitogen, multiply through mitosis which is being treated as proliferation of $CD4^+T$ cells. On that outlook, we consider here the logistic growth of $CD4^+T$ cells in our three component mathematical models in deterministic version but to find out the expected time to extinction of the disease HIV/AIDS, stochastic approach is needed. Thus, we introduce hypothesized transition rates for the stochastic model and the quasi-stationary distribution of the infected CD4$^+$T cells. Our results reveal that proliferation of the CD4$^+$T cells contributes significantly for estimation of the expected time to extinction of the disease through stochastic version.

8.3.1 The Deterministic Model

To generate the deterministic model of HIV infection, we need to consider the population dynamics of CD4$^+$T cells, when HIV enters in to the body. T cells are produced from precursors in the bone marrow and thymus at a constant rate s and have a death rate μ. In presence of HIV virus, antigen or mitogen stimulates T cells. At the time of stimulation, T cells multiply through mitosis with a rate r, i.e., the growth or proliferation of T cells from existing T cells are governed by logistic fashion [4–6]. Here the model is discussed based on the HIV infection on CD4$^+$T cells. The growth rate of the uninfected T cell can be written as,

$$\frac{dx_1}{dt} = s + rx_1\left(1 - \frac{x_1}{T_m}\right) - \mu x_1,$$

where r is the average growth rate of the T cells, the density of uninfected T cells in the blood is x_1. Here, r depends on the degree of idiotypic network stimulation of the T cell proliferation. The total number of T cell is bounded, as T_m is the maximum number of T cell in the body [4, 7]. The dynamics of the HIV infection will be determined by the interactions between the uninfected T cells, infected T cells, and the free virus. There are several research works, where elaborate description of the dynamics of the interactions of T cells and HIV have been discussed.

The interaction of T cells and HIV can be modeled as follows:

$$\frac{dx_1}{dt} = s + rx_1\left(1 - \frac{x_1}{T_m}\right) - \mu x_1 - \beta x_1 v,$$

$$\frac{dx_2}{dt} = \beta x_1 v - \alpha x_2,$$

$$\frac{dv}{dt} = cx_2 - \gamma v. \tag{8.33}$$

Here the density of uninfected T- cells is $x_1(t)$, infected such cells is $x_2(t)$, and free virus is $v(t)$. The infection rate of T cells is 'β', per capita rate of disappearance of infected cells is 'α', rate of production of virus by an infected cell is 'c', and the death rate of the free virus particle is denoted by 'γ'.

We have assumed that at the equilibrium state $\dot{v} = 0$, v can be eliminated by putting $v = cx_2/\gamma$. So x_1 and x_2 satisfy,

$$\frac{dx_1}{dt} = s + rx_1\left(1 - \frac{x_1}{T_m}\right) - \mu x_1 - kx_1 x_2,$$

$$\frac{dx_2}{dt} = kx_1 x_2 - \alpha x_2. \tag{8.34}$$

Here, $k = \beta c/\gamma$, all the parameters and variables are non negative.

8.3.2 The Stochastic Model Formulation

Here we consider two state variables, $x_1(t)$ is the number of uninfected T cells and $x_2(t)$ is the number of infected such cells at time t. They jointly take values in the state space $S = \{(m, n) : m = 0, 1, 2, \ldots; n = 0, 1, 2, \ldots\}$. So, $p_{m,n}(t) = P\{x_1(t) = m, x_2(t) = n\}$ is the joint distribution of $x_1(t), x_2(t)$ at time t. If m and/or n are negative, $p_{m,n}(t)$ is then equal to 0.

The model is based on the following four basic events, i.e., production of CD4$^+$T cells from primary and secondary sources, infection of CD4$^+$T cells, death of CD4$^+$T cells, and death of such infected cells. Here we assume that the total population is N. Table 8.3 represents the hypothesized transition rates of the model (8.34).

8.3.3 Description of the Transition States

Here we have introduced two forms of generation of uninfected T cells. Primary source is from precursors in the bone marrow and thymus and secondary source is proliferation of infected T cells considering that total number of T cell proliferates to its maximum T_m, at which its proliferation shuts off (Tables 8.4 and 8.5).

Table 8.3 Hypothesized transition rates for the stochastic version

Event (CD4$^+$T cells)	Transition	Transition rate
Birth of an uninfected cell	$(m, n) \rightarrow (m + 1, n)$	$\lambda_1(m, n) = s + $ $m(a_1 - a_2 m) + \rho m$
Death of an uninfected cell	$(m, n) \rightarrow (m - 1, n)$	$\mu_1(m, n)$ $= m(b_1 + b_2 m + d_1)$
Infection of an uninfected cell	$(m, n) \rightarrow (m - 1, n + 1)$	$\lambda_2(m, n) = kmn$
Recovery or death of infected cell	$(m, n) \rightarrow (m, n - 1)$	$\mu_2(m, n) = (\rho + \delta)n$

Table 8.4 A hypothetical set of parameter values

Parameter	Value
s	$5 \, \text{day}^{-1} \, \text{mm}^{-3}$ [1]
d_1	$0.01 \, \text{day}^{-1}$ [1]
a	$0.5 \, \text{day}^{-1}$ [1]
a_1	0.30 [8]
b_1	0.015 [8]
a_2	0.02 [8]
b_2	0.001 [8]
β	$0.0002 \, \text{mm}^{-3}$ [1]
k	$\beta q / c$ [1]
ρ	$0.01 \, \text{day}^{-1}$ [1]
δ	$1 \, \text{day}^{-1}$ [1]
q	$800 \, \text{day}^{-1} \, \text{mm}^{-3}$ [1]
c	$5 \, \text{day}^{-1}$ [1]

Table 8.5 Hypothesized transition rates for the stochastic version

Event (CD4$^+$T cells)	Transition	Transition rate
Birth of an uninfected cell	$(m, n) \rightarrow (m + 1, n)$	$\lambda_1(m, n) = s + r x_1$ $\left(1 - \frac{x_1}{T_m}\right)$
Death of an uninfected cell	$(m, n) \rightarrow (m - 1, n)$	$\mu_1(m, n) = \mu m$
Infection of an uninfected cell	$(m, n) \rightarrow (m - 1, n + 1)$	$\lambda_2(m, n) = kmn$
Death of infected cell	$(m, n) \rightarrow (m, n - 1)$	$\mu_2(m, n) = \alpha n$

We also consider here that natural birth of the uninfected T cells or stimulation by antigen or mitogen is balanced by the natural death of uninfected T cells as the programmed cell cycle that continuously operates the process at cellular level to maintain equilibrium state. This cycle continues until or unless it is hindered by any biological or environmental stress factor.

8.3.4 Kolmogorov's Forward Equation

The stochastic model analog can be resolved as the previous sections reveal the analytical techniques applicable in the scenario. The transition states of the model system can be solved by applying the Kolmogorov forward equation. Similarly the time to extinction of the infected cells as well as their distribution is quite an identical process and to avoid we summarily consider the methods with advancement to the numerical results straightway.

Let us assume that in a infinitesimally small time interval Δt, the probability of exactly one birth (or one death) is given by {birth rate(or death rate) $\times (\Delta t) + O(\Delta t)$}, and that of more than one event (birth and/ or death) is $O(\Delta t)$.

Let us consider the probability $p_{m,n}(t + \Delta t)$, where $\Delta t \to 0$. So,

$$
\begin{aligned}
p_{m,n}(t + \Delta t) = {} & p_{m-1,n}(t)\lambda_1(m - 1, n)\Delta t + p_{m+1,n}(t)\mu_1(m + 1, n)\Delta t \\
& + p_{m+1,n-1}(t)\lambda_2(m + 1, n - 1)\Delta t + p_{m,n+1}(t)\mu_2(m, n + 1)\Delta t \\
& + p_{m,n}(t)(1 - K(m, n)\Delta t) + O(\Delta t),
\end{aligned}
$$

where $K(m, n) = \lambda_1(m, n) + \mu_1(m, n) + \lambda_2(m, n) + \mu_2(m, n)$.

It is considered here that events consisting of more than one birth or more than one death are included in the $O(\Delta t)$ term.

Now,

$$
p'_{m,n}(t) = \lim_{\Delta t \to 0} \frac{p_{m,n}(t + \Delta t) - p_{m,n}(t)}{\Delta t}.
$$

Therefore, the Kolmogorov's forward equation for the model can be written as,

$$
\begin{aligned}
p'_{m,n}(t) = {} & p_{m-1,n}(t)\lambda_1(m - 1, n) + p_{m+1,n}(t)\mu_1(m + 1, n) + p_{m+1,n-1}(t) \\
& \lambda_2(m + 1, n - 1) + p_{m,n+1}(t)\mu_2(m, n + 1) - p_{m,n}(t)K(m, n). \quad (8.35)
\end{aligned}
$$

8.3.5 Time to Extinction of Infected T Cells

The time to extinction is an important measure, which can be elucidated through the epidemiological perceptive. It is used to define thresholds for stochastic endemic models [9].

Here the time to extinction of the infected T cells is denoted by the random variable τ, i.e., when we have $\{t \le \tau\}$, infected T cells will exist. Now

$$
P\left(t \le \tau\right) = P\{x_2(t) > 0\},
$$

and

$$
P\left(\tau \le t\right) = P\{x_2(t) = 0\} = p_{.0}(t),
$$

where

$$
p_{.0}(t) = \sum_{m=0}^{N} p_{mn}(t).
$$

Let the c.d.f. and p.d.f. of τ are denoted by F and f respectively.

$$F(t) = P\{x_2(t) > 0 = \sum_{m=0}^{N} P\{x_1(t) = m, x_2(t) = 0\}\} = \sum_{m=0}^{N} p_{m,0}(t)$$

$$f(t) = \sum_{m=0}^{N} p'_{m,0}(t) = p'_{\cdot 0}(t).$$

Putting $n = 0$ in Eq. (8.35) we have,

$$p'_{m,0}(t) = p_{m-1,0}(t)\lambda_1(m-1,0) + p_{m+1,0}(t)\mu_1(m+1,0) + p_{m+1,-1}(t)$$
$$\lambda_2(m+1,-1) + p_{m,1}(t)\mu_2(m,1) - p_{m,0}(t)K(m,0),$$
$$= (s + r(m-1)(1 - \frac{m-1}{T_m}))p_{m-1,0}(t) + \mu(m+1)p_{m+1,0}(t)$$
$$+ \alpha p_{m,1}(t) - (s + rm(1 - \frac{m}{T_m}) + \mu m)p_{m,0}(t).$$

$$\Rightarrow \sum_{m=0}^{\infty} p'_{m,0}(t) = s \sum_{m=0}^{\infty} p_{m-1,0}(t) + r \sum_{m=0}^{\infty} p_{m-1,0}(t)(m-1)(1 - \frac{m-1}{T_m})$$

$$+ \mu \sum_{m=0}^{\infty} (m+1)p_{m+1,0}(t) + \alpha \sum_{m=0}^{\infty} p_{m,1}(t) - s \sum_{m=0}^{\infty} p_{m,0}(t)$$

$$- r \sum_{m=0}^{\infty} p_{m,0}(t)m(1 - \frac{m}{T_m}) - \mu \sum_{m=0}^{\infty} m p_{m,0}(t)$$

$$= \alpha \sum_{m=0}^{\infty} p_{m,1}(t) = \alpha \sum_{m=0}^{\infty} P(x_1(t) = m, x_2(t) = 1).$$

$$\text{i.e., } p'_{\cdot 0}(t) = \alpha p_{\cdot 1}(t), \qquad (8.36)$$

$$\Rightarrow f(t) = p'_{\cdot 0}(t) = \alpha P\{x_2(t) = 1\} = \alpha p_{\cdot 1}(t),$$

where

$$p_{\cdot 1}(t) = \sum_{m=0}^{\infty} p_{m,1}(t).$$

To find the expected time to extinction, we need to compute $p_{\cdot 1}(t)$.

8.3.6 The Distribution of the Time to Extinction

To find the absorption of $x_2(t)$ at 0, the distribution of $(x_1(t), x_2(t))$, $\forall t \geq 0$ is necessary but is not possible.

So we look for the process that will give more detailed information regarding eventual absorption in the class $\{(m, 0) : m = 1, 2, \ldots\}$, which is assured in this perspective. Thus, we try to find out the quasi-limiting distribution.

The process $\{x_1(t), x_2(t)\}, t \geq 0$ has a unique conditional distribution (conditioned being not absorbed) $q_{m,n}$, where $q_{m,n} = \lim_{t \to \infty} P\{x_1(t) = m, x_2(t) = n | x_2(t) \neq 0\}$, whatever the distribution of initial state is $(x_1(0), x_2(0))$.

Let,

$$q_{m,n}(t) = P\{x_1(t) = m, x_2(t) = n | x_2(t) \neq 0\},$$

$$= \frac{p_{m,n}(t)}{1 - p_{.0}(t)}, m = 0, 1, 2, \ldots; n = 1, 2, \ldots;$$

$$\Rightarrow q'_{m,n}(t) = \frac{p'_{m,n}(t)}{1 - \sum_{m=0}^{\infty} p_{m,0}(t)} + (-1) \frac{-\sum_{m=0}^{\infty} p'_{m,0}(t)}{(1 - p_{.0}(t))^2} p_{m,n}(t),$$

$$= \frac{p'_{m,n}(t)}{1 - p_{.0}(t)} + \frac{p_{m,n}(t)}{(1 - p_{.0}(t))^2} \alpha p_{.1}(t).$$

If the initial distribution is assumed to be the quasi-stationary distribution, i.e., $p_{m,n}(0) = q_{m,n}$, $\forall m, n$, then the distribution of the time to extinction will be especially simple.

Let τ_Q be the time to extinction of infected T cells. We can write $q'_{m,n}(t)$ as,

$$q'_{m,n}(t) = \frac{p'_{m,n}(t)}{1 - p_{.0}(t)} + \alpha p_{.1}(t) \frac{p_{m,n}(t)}{(1 - p_{.0}(t))}.$$

Also we can write,

$$q'_{m,n}(t) = \frac{p'_{m,n}(t)}{1 - p_{.0}(t)} + \alpha p_{.1}(t) \frac{q_{m,n}(t)}{(1 - p_{.0}(t))}.$$

Since the initial distribution is assumed to be quasi-stationary, i.e., $p_{m,n}(0) = q_{m,n}$,

$$\Rightarrow q'_{m,n}(t) = \frac{p'_{m,n}(t)}{1 - p_{.0}(t)} + \alpha q_{.1}(t) \frac{p_{m,n}(t)}{(1 - p_{.0}(t))}. \qquad (8.37)$$

Putting $q'_{m,n}(t) = 0$ in Eq. (8.37), we get $p'_{m,n}(t) = -\alpha q_{.1} p_{m,n}(t)$.

Thus, we have $p_{m,n}(t) = e^{-\alpha q_{.1}t}$, constant.

At $t = 0$, we have $p_{m,n}(0) = q_{mn}$,

$$\Rightarrow p_{mn}(t) = q_{mn}e^{-\alpha q_{.1}t} \quad \text{and} \quad p_{.1}(t) = q_{.1}e^{-\alpha q_{.1}t}.$$

Using the above form of $p_{.1}(t)$ in Eq. (8.36) we have,

$$f(t) = p_{.0}'(t) = (\alpha q_{.1}) e^{-(\alpha q_{.1})t}, \quad t > 0.$$

Thus the expected time to extinction of the infected T cells has an exponential distribution and is equal to $E(\tau_Q) = \frac{1}{\alpha q_{.1}}$.

From the diffusion approximation, $q_{.1}$ can be estimated; the quasi-stationary distribution is approximated asymptotically by a bivariate normal distribution [10].

8.3.7 Diffusion Approximation

Let us consider the two-dimensional process (8.34),

$$\frac{dx_1}{dt} = s + rx_1\left(1 - \frac{x_1}{T_m}\right) - \mu x_1 - kx_1x_2,$$

$$\frac{dx_2}{dt} = kx_1x_2 - \alpha x_2. \tag{8.38}$$

If the total population N is sufficiently large, the quasi-stationary distribution is approximated by a bivariate normal distribution (Table 8.6).

The critical point of the deterministic model is given by $\widehat{x} = (\widehat{x}_1, \widehat{x}_2)$, where $\widehat{x}_1 = \frac{\alpha}{k}$ and $\widehat{x}_2 = \frac{sk-\alpha\mu}{\alpha k}$.

Now we want to find the mean change $E(\Delta x)$ and the covariance matrix $E\left(\Delta x(\Delta x)^T\right)$ for the time interval Δt, neglecting terms of order $(\Delta t)^2$.

In the time interval t to $t + \Delta t$, the changes in the state variables x_1 and x_2 are denoted by $\Delta(x_1)$ and $\Delta(x_2)$. So,

Table 8.6 Possible changes in the two-population system (8.34) with the probabilities

Change	Probability
$\Delta x_1 = [1, 0]^T$	$p_1 = \left(s + rx_1\left(1 - \frac{x_1}{T_m}\right)\right)\Delta t$
$\Delta x_2 = [-1, 0]^T$	$p_2 = (\mu x_1 + kx_1x_2)\Delta t$
$\Delta x_3 = [0, -1]^T$	$p_3 = \alpha x_2\Delta t$
$\Delta x_4 = [0, 1]^T$	$p_4 = kx_1x_2\Delta t$

$$E\left(\Delta x\right) = E\left(\Delta x_1 \Delta x_2\right) = \sum_{j=1}^{4} p_j \Delta x_j,$$

$$= \left(s + rx_1\left(1 - \frac{x_1}{T_m}\right) - \mu x_1 - kx_1x_2 - \alpha x_2 + kx_1x_2\right)\Delta t + O(\Delta t),$$

$$= b(x)\Delta t + O(\Delta t).$$

The Jacobian matrix of the vector $b(x)$ with respect to x is denoted by $B(x)$ and it is defined by,

$$B(x) = \frac{\partial b(x)}{\partial x} = \begin{pmatrix} r - \frac{2rx_1}{T_m} - \mu - kx_2 & -kx_1 \\ kx_2 & -\alpha + kx_1 \end{pmatrix}.$$

The approximated value of $B(x)$ at the critical point $\widehat{x} = (\widehat{x}_1, \widehat{x}_2)$ is given by,

$$B(\widehat{x}) = \begin{pmatrix} \frac{-sk}{\alpha} - \frac{r\alpha}{kT_m} & -\alpha \\ \frac{1}{\alpha}[sk + r\alpha\left(1 - \frac{\alpha}{kT_m}\right) - \alpha\mu] & 0 \end{pmatrix}.$$

Now,

$$E\left(\Delta x(\Delta x)^T\right) = \sum_{j=1}^{4} p_j \Delta x_j(\Delta x_j)^T = S(x),$$

$$= \begin{pmatrix} 2s + 2rx_1\left(1 - \frac{x_1}{T_m}\right) & 0 \\ 0 & 2\alpha x_2 \end{pmatrix}.$$

At the critical point $\widehat{x}$, the value of $S(x)$ is given by,

$$S(\widehat{x}) = 2\begin{pmatrix} s + \frac{r\alpha}{k}\left(1 - \frac{\alpha}{kT_m}\right) & 0 \\ 0 & s + \frac{r\alpha}{k}\left(1 - \frac{\alpha}{kT_m}\right) - \frac{\alpha\mu}{k} \end{pmatrix}.$$

The process $N^{1/2}\{x(t) - \widehat{x}\}$ is approximated by a bivariate Ornstein–Uhlenbeck process for large N with local drift matrix $B(\widehat{x})$ and local covariance matrix $S(\widehat{x})$. Then the stationary distribution of the Ornstein–Uhlenbeck process approximates the quasi-stationary distribution, which is approximately bivariate normal with mean 0 and covariance matrix Σ. The matrix Σ can be determined with the help of the local drift matrix $B(\widehat{x})$ and local covariance matrix $S(\widehat{x})$ through the relationship $B(\widehat{x})\Sigma + \Sigma B^T(\widehat{x}) = -S(\widehat{x})$, where the superscript T is used to denote the transpose. After solving the above equation we get,

$$\Sigma = \begin{pmatrix} \sigma_1 & \sigma_2 \\ \sigma_2 & \sigma_3 \end{pmatrix}.$$

The above result leads to the conclusion that in quasi-stationarity, the marginal distribution of the number of uninfected T cells and infected T cells are approximately $N(\mu_{x_1}, \sigma_{x_1})$ and $N(\mu_{x_2}, \sigma_{x_2})$ respectively, where

$$\mu_{x_1} = \widehat{x_1}, \sigma_{x_1} = \sqrt{(\sigma_1/N)}, \mu_{x_2} = \widehat{x_2} \text{ and } \sigma_{x_2} = \sqrt{(\sigma_3/N)}.$$

Further, the covariance is approximated by $\sigma_{x_1 x_2} = \frac{-\alpha}{k}$, where $\sigma_1 = \frac{A}{B}$, $\sigma_2 = \frac{-\alpha}{k}$, $\sigma_3 = \frac{1}{\alpha^2} \frac{AC}{B} + \frac{B}{k}$, and $A = s + \frac{r\alpha}{k}\left(1 - \frac{\alpha}{kT_m}\right) + \frac{\alpha^2}{k}$, $B = \frac{sk}{\alpha} + \frac{r\alpha}{kT_m}$ and $C = sk + r\alpha\left(1 - \frac{\alpha}{kT_m}\right) - \mu\alpha$. Note that the parameters should also satisfy $\sigma_1 > 0$ and $\sigma_3 > 0$.

From diffusion approximation, we have concluded that the marginal distribution of the infected T cell population size in quasi-stationarity is approximately normal. Now again we modify the approximating normal distribution by truncation at $1/2$ to achieve consistency with $x_2 \geq 0$. Hence we find that the distribution can be enunciated by the following approximation:

$$q_{.n} \approx \frac{1}{\sigma_{x_2}} \frac{\varphi\left(\frac{n - \mu_{x_2}}{\sigma_{x_2}}\right)}{\phi\left(\frac{\mu_{x_2} - 1/2}{\sigma_{x_2}}\right)}, \tag{8.39}$$

where ϕ and φ are the standard normal c.d.f. and the standard normal p.d.f. respectively.

8.3.8 The Expected Time to Extinction

We find the expected time to extinction ($E(\tau_Q)$) from quasi-stationary distribution. The expected time to extinction is given by,

$$E(\tau_Q) = \frac{1}{\alpha q_{.1}} = \frac{\sigma_{x_2}}{\alpha} \frac{\phi\left(\frac{\mu_{x_2} - 1/2}{\sigma_{x_2}}\right)}{\varphi\left(\frac{1 - \mu_{x_2}}{\sigma_{x_2}}\right)}. \tag{8.40}$$

We have observed that the expected time to extinction is a function of population size N, which is a function when the population size is increasing (Fig. 8.8).

8.3.9 Numerical Illustration

The entire simulation is done using $Matlab^R$ for approximating quasi-stationary distribution and the expected time to extinction. Here we have adopted Monte Carlo simulations techniques. The hypothesized parameter values are given in Table 8.7, though we consider different population size.

From Fig. 8.8, we can say that when the population size is $N = 100$, the quasi-stationary distribution is truncated and positively skewed, i.e., definitely non-normal. But if we increase the population size, the distribution becomes symmetric, i.e., the quasi-stationary distribution is approximately normal in its body (8.33).

The expected time to extinction from quasi-stationarity for various population size is given in Fig. 8.7. Here we notice that the expected time to extinction grows slowly with N up to 1000 after that it grows faster with N. When population size N is small it is natural that the number of infected T cells is also expected to be small and at that time the human immune system can decelerate the rapid growth of infected cells resulting in little time of extinction of the disease. But when population size is high, unsurprisingly the number of infected cells is expected to be elevated and

Table 8.7 A hypothetical set of parameter values for the model system (8.34)

Parameter	Default value assigned (day^{-1})
s	$10\,mm^{-3}$ [11]
r	0.04 [4, 11]
T_m	$1500\,mm^{-3}$ [4]
k	$\beta * c / \gamma$
β	$0.002 \leq \beta \leq 0.5\,mm^{-3}$ [11, 12]
c	50 Virion/CD4$^+$ T cells [13]
γ	2.0 [13]
μ	$0.01\,mm^{-3}$ [14]
α	$0.24\,mm^{-3}$ [11]

Fig. 8.7 The expected time to extinction in transition region is shown as a function of N, all other parameters are same as in Fig. 8.8

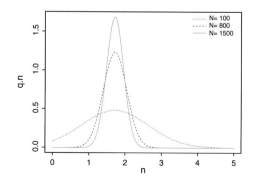

Fig. 8.8 The
quasi-stationary distribution
$q_{.n}$ for different values of N
of Eq. (8.34) for $s = 10$,
$r = 0.04$, $T_m = 1500$,
$\beta = 0.024$, $c = 50$
Virion/CD4$^+$ T cells,
$\gamma = 2.0$, $\mu = 0.01$ and
$\alpha = 0.24$

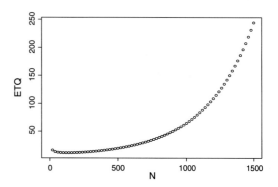

immune system cannot control the rapid growth of infected host cells and ultimately
leads to high extinction time of disease.

Thus it is evident from our numerical analysis that accumulation of infected cells,
which is enhanced when large number of susceptible cells is generated from both
primary and secondary sources. In such a scenario, we find that instantaneous rate of
contact is severely high and contributes higher kurtosis, which implies that the instant
infection recurring probability is much higher in the population. High contact rate
boosts the immunity for larger signal implication, which counter-balance the negative
effect of such alarming contact of infection. Contrastingly, when we consider a low
susceptible population, the contact process gets diluted to a large extent and thereby
the skewness of the population expands (i.e., positively skewed), while the kurtosis
gets a little damped implicating a lower contact rate of the infection (Fig. 8.7).

8.3.10 Discussion and Conclusion

In this section, we have presented a basic mathematical model of HIV infection in
CD4$^+$T cells considering generation of T cells from bone marrow and thymus. But
after invasion of HIV in to the body, antigen or mitogen stimulates T cells and these
healthy cells multiply through mitosis. Based on the concept of quasi-stationarity
and the diffusion approximation, the result can be used to study the transition region.

Introduction of the stochasticity of the existing deterministic model of HIV gives
a positive result in this research work. Approximation of the marginal distribution
through quasi-stationarity has added a new dimension in the epidemiological study.
Our numerical simulation shows that our approximation of the expected time to
extinction is quite satisfactory.

Thus, it can be concluded that the expected time to extinction of the HIV infection from quasi-stationarity is an increasing function of the population size. By this
way, proliferation of T cells through mitosis when HIV invades, has contributed significantly for expected time to extinction, which reveals proper justification by our
analytical and numerical results.

References

1. Zhou, X., Song, X., Shi, X.: A Differential equation model of HIV infection of $CD4^{+}T$ cells with cure rate. J. Math. Anal. Appl. **342**, 1342–1355 (2008)
2. Nowak, M.A., Lloyd, A.L., Vasquez, G.M., Wiltrout, T.A., Wahl, L.M., Bischofberger, N., Williams, J., Kinter, A., Fauci, A.S., Hirsch, V., Lifson, J. D.: Viral dynamics of primary viremia and anti retroviral therapy in simian immunodeficiency virus infection. J. Virol. **71**(10), 7518–7525 (1997)
3. Spouge, J.L., Shrager, R.I., Dimitrov, D. S.: HIV-1 infection kinetics in tissue cultures. Math. Biosci. **138**(1), 1–22 (1996)
4. Perelson, A.S., Nelson, P.W.: Mathematical analysis of HIV-1 dynamics in vivo. SIAM Rev. **41**, 3–41 (1999)
5. Wang, L., Li, M.Y.: Mathematical analysis of the global dynamics of a model for HIV infection. Math. Biosci. **200**, 44–57 (2006)
6. Asquith, B., Bangham, C.R.M.: The dynamics of T cell fratricide; application of a robust approach to mathematical modeling in immunology. J. Theoret. Biol. **222**, 53–69 (2003)
7. Roy, P.K., Chatterjee, A.N.: Electrical engineering and applied computing. In: Sio long Ao (ed.) Effect of HAART on CTL Mediated Immune Cells: An Optimal Control Theoretic Approach, vol. 90, pp. 595–607. Len Gelman Springer, New York (2011)
8. Matis, J.H., Kiffe, T.R.: On the Cumulants of Population Size for the Stochastic Power Law Logistic Model. Theor. Popul. Biol. **53**, 16–29 (1998)
9. Nassel, I.: On the time to extinction in recurrent epidemics. J. R. Statist. Soc. **61**(2), 309–330 (1999)
10. Barbour, A.D.: The principle of the diffusion of arbitrary constants. J. Appl. Probab. **9**, 519–541 (1972)
11. Roy, P.K., Chatterjee, A.N.: T-cell proliferation in a mathematical model of CTL activity through HIV-1 infection. In: Lecture Notes in Engineering and Computer Science: Proceedings of The World Congress on Engineering 2010, WCE 2010, 30 June–2 July, London, U.K., pp. 615–620 (2010)
12. Wodarz, D., Nowak, M.A.: Specific therapy regimes could lead to long-term immunological control to HIV. Proc. Natl. Acad. Sci. USA **96**(25), 14464–14469 (1999)
13. Roy, P.K., Sil, N., Bhatterjee, S.: On the estimation of expected time to extinction in a dynamical model of HIV. Int. J. Math. Sci. Appl. **2**(1), 213–221 (2012)
14. Tuckwell, H.C., Wan, F.Y.M.: Nature of equilibria and effects of drug treatments in some simple viral population dynamical models. IMA J. Math. Appl. Med. Biol. **17**, 311–327 (2000)

Printed in the United States
By Bookmasters